PROFESSIONAL DEVELOPMENT OF SCIENCE TEACHERS

LOCAL INSIGHTS WITH LESSONS FOR THE GLOBAL COMMUNITY

REFERENCE BOOKS IN INTERNATIONAL EDUCATION
Edward R. Beauchamp, Series Editor

PROFESSIONAL DEVELOPMENT OF SCIENCE TEACHERS
LOCAL INSIGHTS WITH LESSONS FOR THE GLOBAL COMMUNITY

EDITED BY
PAMELA FRASER-ABDER

Routledge
Taylor & Francis Group

LONDON AND NEW YORK

First Published 2002 by
RoutledgeFalmer

Published 2013 by Routledge
4 Park Square, Milton Park, Abingdon, Oxon OX14 4RN
605 Third Avenue, New York, NY 10017

Routledge is an imprint of the Taylor & Francis Group, an informa business

Library of Congress Cataloging-in-Publication Data

Fraser-Abder, Pamela, 1948–
    Professional   development   of   science   teachers   :   local   insights   with
    lessons for the global community / Pamela Fraser-Abder.
        p. cm. —
    ISBN 0-8153-3912-7
        1. Science teachers—In-service training. 2. Science—Study and teaching (Continuing edu-
cation) I. Title.

Q181.F8346 2002
507'.1'55—dc21   2001048676
ISBN   978-0-815-33912-0 (hbk)

Contents

Series Editor's Preface

This series of scholarly works in comparative and international education has grown well beyond the initial conception of a collection of reference books. Although retaining its original purpose of providing a resource to scholars, students, and a variety of other professionals who need to understand the role played by education in various societies or world regions, it also strives to provide accurate, relevant, and up-to-date information on a wide variety of selected educational issues, problems, and experiments within an international context.

Contributors to this series are well-known scholars who have devoted their professional lives to the study of their specializations. Without exception these men and women possess an intimate understanding of the subject of their research and writing. Without exception they have studied their subject not only in dusty archives, but have lived and traveled widely in their quest for knowledge. In short, they are "experts" in the best sense of that often overused word.

In our increasingly interdependent world, it is now widely understood that it is a matter of military, economic, and environmental survival that we understand better not only what makes other societies tick, but also how others, be they Japanese, Hungarian, South African, or Chilean, attempt to solve the same kinds of educational problems that we face in North America. As the late George Z. F. Bereday wrote more than three decades ago: "[E]ducation is a mirror held against the face of a people. Nations may put on blustering shows of strength to conceal public weakness, erect grand façades to conceal shabby backyards, and profess peace while secretly arming for conquest, but how they take care of their children tells unerringly who they are" (*Comparative Methods in Education*, New York: Holt, Rinehart and Winston, 1964, p. 5).

Perhaps equally important, however, is the valuable perspective that studying another education system (or its problems) provides us in under-

viii
                                          Series Editor's Preface

standing our own system (or its problems). When we step beyond our own limited experience and our commonly held assumptions about schools and learning in order to look back at our system in contrast to another, we see it in a very different light. To learn, for example, how China or Belgium handles the education of a multilingual society; how the French provide for the funding of public education; or how the Japanese control access to their universities enables us to better understand that there are reasonable alternatives to our own familiar way of doing things. Not that we can borrow directly from other societies. Indeed, educational arrangements are inevitably a reflection of deeply embedded political, economic, and cultural factors that are unique to a particular society. But a conscious recognition that there are other ways of doing things can serve to open our minds and provoke our imaginations in ways that can result in new experiments or approaches that we may not have otherwise considered.

Since this series is intended to be a useful research tool, the editor and contributors welcome suggestions for future volumes, as well as ways in which this series can be improved.

Edward R. Beauchamp
University of Hawaii

Introduction

Science teacher educators entered the twenty-first century with a global call to provide science teachers with professional development that is successful, appropriate, and effective and has the long-term effect of producing global scientific literacy. The goal, as proposed by U.S. national standards and the United Nations, is to achieve scientific literacy for all, a goal that has to be achieved in a world with an increasingly diverse student population with different cultural perspectives, experiences, and expectations of education; different styles of learning and behavior; and teachers and parents who bring differing views of teaching and learning to our schools. The attitudes, skills, and knowledge required by teachers to achieve this new vision for science education is both broad and deep, requiring science teacher educators to reevaluate their models for professional development, since clearly the traditional models for professional development have proven to be inadequate for helping teachers to achieve the goals of scientific literacy for all.

As noted by Loucks-Horsley and colleagues, the current state of professional development includes:

1. Significant numbers of teachers who have few or no professional development opportunities.
2. Development opportunities that may not be appropriate to the learning goals or provide insufficient support over time for teachers to apply what is learned in classrooms.
3. A focus on individual development, one teacher at a time, without attention to organizational development.
4. Some pockets of innovation but minimum means for greater impact both within individual systems and beyond. (Loucks-Horsley, Hewson, Love, & Stiles, 1998)

Designing and implementing meaningful, effective professional development initiatives for science education is complex and fraught with many barriers and pitfalls. A strong knowledge base and a great deal of consensus about what constitutes effective professional development exist; but unfortunately there is a gap between knowledge and practice. According to Loucks-Horsley et al. one reason for this gap is the lack of rich descriptions of effective programs constructed in various contexts addressing common challenges in unique ways.

Purpose of This Book

This book is intended to provide a description of the practices and issues that arise during professional development of science teachers in a global context. It is both a description of science teacher education models and a discussion of the issues and challenges that arise during professional development. It provides a rich description of the diverse contexts and challenges in teaching science and the many barriers to effective professional development of teachers of science.

Professional development in science education is an area that has, to some extent, been investigated in isolated units in specific countries. However, we have not as educators done a comparative analysis of what issues and perspectives arise during professional development for science teachers. The following chapters provide such an analysis using Britain, Canada, China, Germany, Greece, Indonesia, Israel, Italy, Japan, Kenya, the Philippines, Trinidad and Tobago, and North, Central, and South America to contextualize professional development for science teachers. Issues investigated include the cultural, equity, and contextual factors that must be considered during professional development in global communities; the effects of the reform movement and the results of The Third International Mathematics and Science Study (TIMSS) on professional development; and the development of science teacher education models and their inherent challenges. This book serves (1) to expose science teacher educators to a global view of professional development and to some of the research that has been conducted during the development and implementation of these professional development models and (2) to provide science teachers with knowledge about the cultural and educational backgrounds of students in their classrooms.

The book provides an avenue for sharing our successes in professional development and alerting others to the causes of our failures. Each chapter addresses specific issues in one or a group of countries. The authors are science teacher educators who are either residents of the country or have been involved in research studies in these countries. Readers thus get a firsthand look at the professional development of the teachers of the students who participated in the TIMSS.

Reference

Loucks-Horsley, S., Hewson, P. W., Love, N., and Stiles, K. (1998). *Designing professional development for teachers of science and mathematics.* Thousand Oaks, CA: Corwin Press.

Chapter 1
Starting Points for Transformation
Resources to Craft a Philosophy to Guide Professional Development in Elementary Science

BONNIE SHAPIRO AND SANDY LAST

Over the past several years we have worked to develop experiences for pre-service elementary science teachers and practicing teachers in the current climate of intense challenges in elementary education. Sandy, an elementary school administrator and MED graduate student, has worked to build a program in her school to support teachers' work in elementary science teaching. Bonnie, a university teacher and researcher, situates her work in a program of teaching and research that gives high status to the knowledge teachers bring to and integrate into their professional development experiences. For many years, our discussions and professional collaborations have dealt with helping preservice and inservice educators understand the complex nature of the work of teaching elementary science. We have explored features such as teacher knowledge and action; the construction of meaning in science; the nature of environments constructed for learning; the implications of provincial, local, national, and international curriculum policies; and social-cultural issues. More recently, our concerns have focused on the importance of teacher support in professional development. In our own provincial school jurisdiction in western Canada, a new Program of Studies in Elementary Science has been implemented, yet little support is provided for teachers to help them understand and embody the approaches to practice mandated in the curriculum. Some of these challenges are new ways of understanding learning, the meaning of inquiry in science teaching, and technological problem solving.

This chapter emerges from our discussions about articulating foundational ideas to craft a philosophy to guide efforts to help teachers through the provision of professional development experiences in science education. The chapter provides a summary of some of our work to locate resources to craft a philosophy to guide professional development that gives high status to teachers' efforts to link their strong experiential knowledge of classrooms and ways of working with students to the task of learning new disciplinary knowledge and educational theories. In this chapter, we present the results of our search for resources and some of the larger organizing themes, insights, and resources we have found useful, which may be of help to others engaged in similar work.

Insights from the Literature: Professional Development and Science Teaching

A review of the research on the professional development of science teachers shows that until recently "there has been little rapprochement between the science educator and staff development communities" (Marx, Freeman, Krajcik, & Blumenfeld, 1998, p. 668). In fact, until recently there has been little attention given to professional development in the educational literature overall. In the past, those writing on teacher development tended to refer less to the work of teaching in subject area disciplines and more to issues that cross subject matter lines, such as classroom management and school effectiveness issues (Marx et al., 1998). In more recent years, greater attention has been given to the contributions, needs, and knowledge of teachers themselves in considering professional development opportunities. This is seen in Tobin, Tippins, and Gallard's (1994) *Handbook of Research on Science Teaching and Learning.* The authors discuss two factors essential for successful professional development: (1) The importance of connecting professional development efforts to teachers' previous knowledge base and (2) the need to provide a supportive, long-term environment for change. Other current writers in the field reiterate these themes and add other useful ideas that help shape thinking about creating professional development opportunities for science teachers.

In *Teaching as the Learning Profession,* editors Darling-Hammond and Sykes (1999) argue that teacher learning is critical to school reform and to the ultimate goal of improving student learning. Writing in this volume, Hawley and Valli (1999) describe four recent research advances that influence thinking about the design of professional development for teachers:

1. Research on school improvement that links change to professional development.
2. Growing agreement that students should be expected to achieve much higher standards of performance, which include a capacity for complex and collaborative problem solving.
3. Research on learning and teaching that reaches substantially different conclusions about how people learn from those that have shaped contemporary strategies for instruction and assessment.

4. Research that confirms the widespread belief among educators that conventional strategies for professional development are ineffective and wasteful and that provides support for the adoption of different ways to facilitate professional learning.

Hawley and Valli call for more opportunities to learn with colleagues that are linked to solving authentic problems defined by the gaps between goals for student achievement and actual student performance (p. 127). They recommend that the following concepts be included in thinking about professional development opportunities:

1. *Goals and Student Performance*—Professional development should be driven by analyses of the differences between goals and standards for student learning and student performance.
2. *Teacher Involvement*—professional development should involve learners (such as teachers) in the identification of what they need to learn and, when possible, in the development of the learning opportunity and the process to be used.
3. *School Based*—professional development should be primarily school based and integral to school operations.
4. *Collaborative Problem Solving*—professional development should provide learning opportunities that relate to individual needs but for the most part are organized around collaborative problem solving.
5. *Continuous and Supported*—professional development should be continuous and ongoing, involving follow-up and support for further learning, including support from sources external to the school that can provide necessary resources and an outside perspective.
6. *Information Rich*—professional development should incorporate evaluation of multiple sources of information on outcomes for students
7. *Theoretical Understanding*—professional development should provide opportunities to engage in developing a theoretical understanding of the knowledge and skills to be learned.
8. *Part of a Comprehensive Change Process*—professional development should be integrated with a comprehensive change process that deals with impediments to and facilitators of student learning. (p. 136)

Loucks-Horsley, Hewson, Love, and Stiles (1998) propose a set of seven principles from their review of the discipline-specific perspectives of professional organizations such as the National Center for Improving Science Education, the National Council of Teachers of Mathematics, and the National Research Council. They blend these ideas with the conclusions of organizations involved more specifically with staff development in their book *Designing Professional Development for Teachers of Science and Mathematics*. This reference, "intended to bring together in one place a thick description and rich discussion of the practices and issues of professional development for mathematics and science education," and provides the following guidelines:

1. Effective professional development experiences are driven by a well-defined image of effective classroom learning and teaching—for example, commitment to all children learning mathematics and science; an emphasis on inquiry-based learning, investigations, problem solving, and applications of knowledge; an approach that emphasizes in-depth understanding of core concepts and challenges students to construct new understandings; and clear means to measure meaningful achievement.

2. Effective professional development experiences provide opportunities for teachers to build their knowledge and skills. For example, they help teachers develop in-depth knowledge of their disciplines as well as pedagogical content knowledge (listening to students' ideas, posing questions, and recognizing common and naïve misconceptions), and they help in choosing and integrating curriculum and learning experiences.

3. Effective professional development experiences use or model with teachers the strategies teachers will use with their students. For example, they start where teachers are and build from there; provide ample time for in-depth investigations, collaborative work, and reflection; and connect explicitly with teachers' other professional development experiences and activities.

4. Effective professional development experiences build a learning community—for example, continuous learning is a part of the school norms and culture, teachers are rewarded and encouraged to take risks and learn, and teachers learn and share together.

5. Effective professional development experiences support teachers to serve in leadership roles, for example, as supporters of other teachers, as agents of change, and as promoters of reform.

6. Effective professional development experiences provide links to other parts of the educational system. For example, professional development is integrated with other district or school initiatives or district or state curriculum frameworks and assessments or both, and it has active supports within the community.

7. Effective professional development experiences are continuously assessing themselves and making improvements to ensure positive impact on teacher effectiveness, student learning, leadership, and the school community. (p. 36)

The National Science Resources Center (1997), in *Science for All Children,* a series of strategies to create innovative professional development programs:

1. Provide continuous and sustained support for professional development—support from school administration must go beyond rhetoric and take the form of stressing science as a basic in the school curriculum.
2. Provide teachers with time to engage in professional development activities—time during the school should be given to participate in professional learning work.
3. Create an environment of collegiality and collaboration—strong professional relationships enable teachers to feel comfortable sharing ideas, acknowledging difficulties, and solving problems that they encounter in the classroom.
4. Incorporate the change process into the professional development design—allow for the "growing pains" of new ways of teaching, learning, and curriculum design. (pp. 78–81)

Finally, Stein, Smith, and Silver (1999) describe what they call a new, more transformative paradigm for professional development. They point out that teachers will need to relearn aspects of their practice to counter their own prior learning experiences, suggesting that the following needs must be met in this effort:

- Teachers need assistance that focuses on their day-to-day efforts to teach in new and demanding ways. Assistance must be embedded in or directly related to individual, daily practice.
- Teacher assistance must be grounded in the content of teaching and learning. Calls for reform require meeting high standards of content knowledge in the disciplines, and teachers must be provided with experiences that support their content understandings.
- Teachers should have access to communities of professional practice to counter the isolation, which is typical of teaching. Learning to work with small to larger groups of teaching colleagues is a component of the professional learning that needs to occur.
- Collaboration with experts outside the teaching community is required to bring different kinds of expertise to the table around the problems that practice may pose.
- Organizational context must be considered. The multiple contexts that influence teachers (e.g., school culture, community, district initiative, provincial curriculum) should be considered as to their constraints and alternatives offered. (p. 241)

The key themes in these recommendations are the need for continuous, long-term support for teacher efforts and a recognition of teachers' significant contributions to their own development. Still, one afternoon workshop designed to transmit information is the most common professional development experience for teachers in school settings. Kennedy (1999) provides a strong critique of the one-shot workshop for professional development in science. She states, "This event has been criticized by virtually every teacher who has ever participated in it and by virtually everyone else even vaguely interested in improving teaching" (p. 1). Kennedy cites the recommendations of researchers and policy analysts who argue that continuing education programs for teachers should be lengthy, that teachers should have a role in defining content, that meetings and learning should be interspersed with classroom practice rather than concentrated into a short periods of time, and that teachers should work in groups rather than in isolation. She makes an additional strong case for an emphasis on the content of professional development programs. In a review of 93 studies, she shows that the more successful professional development programs do not simply emphasize discipline knowledge, but instead are about what to teach and about how students learn subject matter. Her central message is to attend to content of professional development first, before form and structure. She also advocates an emphasis on treating teachers as the professionals they are, by allowing them to discover what works best rather than imposing prescriptive approaches that give little practical leeway. One of the areas teachers themselves identify as a significant professional development topic is their interest in acknowledging the ideas and understandings students bring to learning. Teachers seek new insights for building forms of practice that help learners construct, rather than receive, transmitted knowledge.

Building New Ideas about the Nature of Meaning Construction in Science and Science Learning

Teachers' views about the nature of knowledge, the nature of science, and the nature of learning impact the ways that they represent science to learners and the approaches they take to organize resources and experiences for science learning. Despite developments in thinking about knowledge creation in science and new understandings about how meaning is constructed by learners, the predominant approaches to teaching that persist are based on the familiar conduit metaphor of knowledge. This metaphor, rooted in the view that knowledge is fixed, is found "out there" in the world and is transferred from the scientist to the teacher to the student. As Duschl (2000) writes, the traditional approach to interpreting science curriculum resources and practices has been one that asks, "What is it that we want students to *know* and what do they need to *do* to know *it*?" (p. 187) The ultimate goal in a program of study based on this perspective is that students acquire scientists' ideas. Duschl points out that when we teach these ideas, that is, what is known in science, without building deep insight into the ways we come to know, we eliminate the chance for learners to grasp deeper understanding of the processes of science, the "social, cognitive and epistemic dynamics that make science an objective way of knowing." (p. 187).

Constructivist thinking about the creation of knowledge in science is a useful framework for developing professional learning experiences for teachers. Bencze and Hodson (1999) assert that while such views of the nature of the scientific endeavor have been widely accepted by the scientific and academic communities, many teachers and many school science curricula continue to promote a view of scientific practice that is locked in a philosophical mind-set that predates these constructs. The following are among the several myths about scientific inquiry that are still very much in evidence in school science curricula:

- Observation provides direct and reliable access to secure knowledge.
- Science starts with observation.
- Science proceeds via induction.
- Experiments are decisive.
- Science comprises discrete, generic processes.
- Scientific inquiry is a simple, algorithmic procedure.
- Science is a value-free activity. (p. 522)

We are attempting to build a philosophy of working with learners and teachers to enhance considerations of the processes of knowing about science and also about the personal meaning of science learning for the student. To build insights into these areas, we have sought resources that value and acknowledge the interpretive nature of understanding that we might use to build preservice and inservice professional development opportunities to give experiences and insight into constructivist perspectives on science, learning, and teaching. Last (2001) lists several resources that have been found to be useful in providing professional development experiences in science to colleagues. She works with other administrators to give participants extended opportunities to engage in dialogue and discussion and to begin to apply new ideas and thinking about science learning in their classrooms. Brooks and Brooks' (1999) volume, *In search of understanding: The case for constructivist classrooms,* has been very helpful in clarifying the philosophical foundations guiding constructivist practice in today's classrooms. Shapiro (1996) describes a project to help preservice teachers develop deeper insights to the process of the creation of objective knowledge in science. Preservice elementary student teachers were guided to design their own data collection projects and grappled with the challenges of identifying variables and collecting information to provide new knowledge about questions they had posed themselves. Although focusing on simple questions, students were surprised by the many unexpected design challenges they posed. Despite their initial discomfort, in the end students expressed satisfaction with the depth of new insight this experience provided into the social, cognitive and epistemic understandings about science in creating new knowledge, which Duschl (2000) asserts must be made more explicit in science learning settings. The Annenberg Project Materials programs, such as Looking at Learning . . . Again (Smithsonian Institution, 2000), provide many excellent starting points for discussion through video case examples that present teachers discussing these issues and their practice in their own classroom settings. Another resource that has been useful in developing understanding of new perspectives on knowledge construction is Shapiro's (1994) study of children learning using a constructivist perspective. The work first attempts to clarify the history and meaning of the constructivist perspective:

- Science knowledge is not "waiting to be discovered," it is we who create meanings.
- Knowledge is a construction.
- Knowledge is socially constructed.
- We hold commitments not only to ideas but also to entire schemes of understanding. (p. 5)

The second half of the book contains case studies of children learning about the topic of light in their grade five classroom. Individual stories of knowledge construction are provided, and the personal meaning students attribute to science learning is presented. Like Duschl, Shapiro asserts that learning science is not just about *knowledge acquisition*. In addition, this work asserts that although science learning is about *concept change*, that is not all it entails. What is needed is a significant emphasis on the nature of *knowledge construction*. The implications of each goal, the view of the learner, and the role of the teacher are outlined in three trends in thinking about science teaching, knowledge acquisition, concept change, and knowledge construction, as shown in Table 1.1.

Classic and foundational studies have been useful in helping preservice and inservice educators understand the development of thinking about science knowledge and its connection to teaching approaches. Thomas Kuhn's (1962) groundbreaking work has been useful in showing that scientific knowledge is the result of consensus among members of a scientific community at a given point in time, inspiring views of classrooms as knowledge-building communities. Another author valued by teachers, who writes about the philosophical foundations of constructivist thinking in

Table 1.1.

Toward a Constructivist Conception of Science Learning and Teaching

	Goal of Science Teaching	View of the Learner	View of the Teaching Task
Trend I	*Idea acquisition*	*"Tabula rasa"* The learner's mind is a blank slate.	*Transmission* Transmit scientists' ideas about natural phenomena to the learner.
Trend II	*Idea change*	*Misinformed* The learner has ideas about the nature of phenomena (misconceptions that need correction).	*Correction* Replace learner's ideas with scientific ideas.
Trend III	*Knowledge construction* Development of the ability to grasp scientific explanations and approaches to understanding events and phenomena.	*Active Participant in Learning* The learner has ideas about the nature of phenomena (alternate frameworks). Each individual also holds a personal view about the purpose and value of many features of the school science learning experience. Learners enter classrooms already seeking to make sense of the world. Each person has a personal approach to understanding. The meanings constructed by learners during science lessons may be very different from those intended by teachers and curriculum authors. In order to grasp new explanations, learners must become aware of the value of considering things in new ways: scientific explanations and approaches to science learning. The learner must actively participate in this personal construction of new meaning.	*Guiding Construction* Assist the learner in the active consideration and construction of ideas by providing opportunities. Become familiar with pattern in students' current understandings and experiences with phenomena available in research literature. Become familiar with pattern of knowledge among one's own students. Provide opportunities to assist learners in their own active construction of meaning by using such approaches as assisting learners in the identification of meaningful problems and organizing learning experiences that emphasize extensive use of language in the creation of meaning, such as journal writing, large and small group discussion, and the use of knowledge organization tools.

Shapiro (1994), p. 198.

mathematics education, is Ernest Von Glasersfeld (1988). He describes the
goals of construction:

> Knowledge cannot aim at a "truth" in the traditional sense by concern for
> the construction of paths of action and thinking that an unfathomable "real-
> ity" leaves open for us to tread. The test of knowledge, therefore, is not
> whether or not it accurately matches the world as it might be "in itself"—a
> match which, as the skeptics have reiterated, we could never check out—
> but whether or not it fits the pursuit of our goals, which are always goals
> within the confines of our own experiential world. (p. 7)

These ideas link to form a view of knowledge that involves ongoing
interpretation of a reality that exists within the human observer and that
asserts that the purpose of meaning construction is not the discovery of reality, but rather the development of ways to better cope with it. Vygotsky's
work (1978) has provided valuable philosophical foundations in the teaching
of many subject areas. Vygotsky's work integrates ideas about the use of language and development and highlights the social nature of knowledge growth
within discipline communities and learning cultures. Novak and Gowin's
(1985) excellent resource, *Learning How to Learn*, helps teachers build foundational understanding of issues in knowledge construction and contains
suggestions for teachers to help learners take greater responsibility for their
own learning.

Cultural Constructions and Science Learning:
The Environments We Create for Learning

Semiotic studies of science learning show how the signs and symbols of our
environment speak messages to those who inhabit learning places. Signs and
symbols are not restricted to the posters on bulletin boards, but include cultural practices such as the habitual patterns we use to interact with learners.
Messages are sent about what science is and how it proceeds in unexpected
ways, including the ways that architecture is organized and presented to
inhabitants of buildings—what sort of space do we attribute to science?—
and the ways teaching resources and materials are presented—what kinds of
people are portrayed as pursuing careers in science? Messages are even conveyed through the kinds of odors and colors that students are exposed to in
learning settings. These become "objects of meaning" (Shapiro, 2000) that
speak powerful messages about what it means to do science, to be a scientist,
and to learn science. We build on resources describing semiotic studies in
science learning (Cunningham, 1984, 1987; Groisman, Shapiro, & Willinsky,
1991; Lemke, 1984, 1990; Shapiro, 1998; 2000; Shapiro & Kirby, 1998) and
the territories we create for learning.

Danesi (1994) writes that cultures are social territories as well as ideological entities. As the cultural elaboration of shelter, architecture, for example, is "the art of imbuing living spaces with symbolic meaning" (p. 186). School settings convey elaborate systems of signs to learners indicating what it means to be a competent learner in the science classroom. We reward students according to the degree that they model these signs for us. Children coming from a particular social class or who speak a specific language are at an advantage in such a system. Lesson organization, language use, and activity patterns are largely structured for students by teachers using unexamined cultural values, sometimes only by representatives of one gender or race, presenting a significant message of exclusion to others. Language and dialogue structures may also become ritualized in school settings. We need to take notice of the ways that communication structures permit students to use scientific language and thinking for themselves and how these structures influence their intellectual development and opportunities to discover and challenge new ideas. The ways we organize time and the amount of time we devote to working with a particular topic are other ways that we speak a message to students about the importance of Science. (Shapiro, Richards, Ross, & Kendal-Knitter, 1999).

Participating in semiotic examinations about what we are doing in science education allows us to see freshly how learners may emerge from the classroom with systems of signification that may not be what we intend. For example, the rigid presentations of structures such as the scientific method or learning the specific steps in technological problem solving may become, for many teachers and students, the embodiment of science itself, creating an image of science as a prescriptive technique rather than a complex process of coming to know. Likewise, science is often portrayed as a game of term memorization or may be seen to focus on ideas that must simply be accepted rather than deeply understood. With such portrayals, learners begin to see themselves as deficient when confronted with very difficult ideas about science content, and they learn not to ask questions. The use of such signs of what science is all about can inhibit rather than facilitate learning experiences in the classroom.

Cobern's (1998) edited volume of papers on cultural perspectives on science education is an excellent source to help articulate thinking about such culturally constructed features of science education as gender, issues of critical reform, equity and language, and second language teaching and science education. Cobern's view is that science is an aspect of culture, and, therefore, it is appropriate to clarify the cultural context of Western science education and how it differs from other contexts. This is important because the beliefs individuals bring to class are supported by culture, and "science education is successful only to the extent that science can find a niche in the cognitive and cultural milieu of students" (p. 8).

Cunningham (1984, 1987) comments on the value of helping teachers become more aware of the process of semiosis or sign making and interpreting capacities operating within the students we teach. He considers the job of nurturing the understanding of sign making processes along with the subject matter being taught. This view of knowledge directs the attention of teachers away from teaching only specific bits of knowledge and toward the cultivation of higher intellectual skills.

As Aikenhead (2000) notes, many studies dealing with school science learning culture have focused on the transmission of scientific information or giving direct information concerning the culture and body of science. Enculturation into Western science typically supports only those students (potential scientists) whose cultural identities harmonize with the culture of Western science (p. 246). Aikenhead urges recognizing that Western science is itself a subculture, and he goes beyond this idea to argue for the need to treat learning as a culture-making process that engages students with who they are and where they are going. Aikenhead's position on culture differs from the social constructivist view of enculturation in which learners are supported in using scientific ways of knowing through social processes, making personal sense of scientific representations of phenomena in terms of their existing everyday knowledge (p. 247).

Issues of Significance: The Need for Warranted
Activities in Professional Development

The purpose of professional development is to bring about positive changes in teaching and learning. As Richardson (1990) notes, teacher change is a necessary condition for system change. Teachers change their ideas and practices regularly, so the problem is usually not that change does or does not happen; rather, there is concern with the degree to which teachers are able to engage in the dialogue about what they consider warranted change and new practice (p. 14). If teachers are to employ research findings and new innovative practices, they must be embedded with theoretical frameworks that are of importance to them. This requires understanding teachers' own frameworks of classrooms and talking about how the premises of a proposed change agree or disagree with one's own premises. Empowerment is threatened when teachers are asked to make changes in activities without being invited to examine their own frameworks or to express their ideas about appropriate fit. Marx et al. (1998) refer to the generative and constructive nature of this process for planning for innovation by referring to the process whereby teachers integrate innovation into classroom practice with their own theoretical views, *enactment*. If new practices are to be accepted and enacted by teachers, they must be considered warranted practices. Connected to this is the importance of authenticity of practice. One approach to engaging teachers in warranted, authentic practice is action research.

Authentic science, science connoisseurship, and the teacher as a transformative curriculum maker are goals that Bencze and Hodson (1999) maintain can be met through action research experiences in professional development in science. Viewing teachers as curriculum makers, they point out that teachers must be supported in their efforts to create a curriculum in science that has an inquiry orientation. They suggest that a supportive but critical environment, one where teachers work closely with a researcher or a "critical friend," is one way to create change in teacher practice and student learning. They maintain,

> If critical thinking, creativity, and skillful problem solving are to be developed by students, it is essential that those who are responsible for that education also possess these attributes. If we want students to acquire the capacity to work productively in a collaborative mode, their teachers must have this capacity, too. It is unrealistic to expect students to have confidence in their own knowledge, skills, and judgement (i.e., to be intellectually independent) if their teachers have been socialized into blind acceptance of the views and decisions of others. (p. 524)

Action research projects can provide resources for renewal, extension of professional confidence, and a sense of teacher empowerment that will ensure continued learning.

Rosebery and Puttick (1992) describe teacher professional development as "situated inquiry." They emphasize the importance of acquiring a discourse of science for teacher development. Acquiring this discourse means participating in scientific inquiry as well as inquiry into teaching. Teachers can do this through participation in such activities as (1) engaging in debate with colleagues about ideas; (2) asking and seeking answer to questions; (3) learning from studying the work of others; (4) making sense of scientific descriptions and explanations; (5) messing about with problems and materials; (6) successively refining experimental apparatuses and collecting, analyzing, and interpreting data; (7) constructing and interpreting graphical representations, comparing methods and results to those of others; and (8) using the theories of others including the standard explanations of science as tools in teaching work (p. 24).

The resources we have selected are designed to bring us closer to the contributions of teachers themselves. We believe that achieving a deeper understanding of the situated nature of change for teachers is best gained by looking to teachers themselves as resources. In fact, reviewing one another's stories of practice can itself initiate the process of change. When well crafted, such stories can capture deep insights into the features of excellent practice. One teacher's story can inspire another. Examples of case studies of innovative practice include those in Bentley and Watts' (1989) edited volume, *Learning and Teaching in School Science: Practical Alternatives.* Polman's (2000) narrative of a teacher designing project-based science experiences is another excellent example of the growing literature in this area.

Following another line of work and resources for teachers inspiring teachers, Tobin, Roth, and Zimmerman (2001) develop an intimate approach for learning about teaching called "coteaching." Although the practice emerges from a preservice setting, the benefits of working and learning alongside a fellow colleague, say, a teacher or administrator, creates the opportunity for a "cogenerative dialogue" about the experience that opens up new understandings and feelings about teaching. Duckworth (1997) extends this type of collegial engagement into large group investigation and collaboration in research on teaching as teachers share and discuss with one another their stories of diverse approaches to similar teaching challenges.

The Heart of Professional Development: Reflection with Self and Others

Research and thinking on the teacher's personal engagement in professional development shows how teachers themselves are the most important resource of the professional development experience. Based on work in their three-year study of New Zealand teachers, Bell (1998) and Bell and Gilbert (1996) describe a model of teacher development as "professional, social and personal development as well as learning," with teacher development considered a form of human development (Bell, 1998, p. 681). Bell and Gilbert see *professional development* as typically taking place in a formally structured setting, as an opportunity to learn to use different teaching activities, work on a curriculum, or examine personally and socially constructed beliefs and ideas about science education. *Social development* involves a construction and reconstruction of what it means to be a teacher often in association with other educators. Hence social development is enhanced by working with others and through the development of social structures and opportunities that permit interaction, discussion, and reflection on these meanings. Personal development involves the individual construction of meaning about what it means to be a teacher and the evaluation for oneself of socially constructed ideas about what it means to be a teacher. Personal development also involves "managing the feelings associated with changing one's activities and beliefs about science education." Bell and Gilbert stress that to address the personal in teacher development means that we must address the creation of social activities that allow social negotiation and reconstruction of ideas about what it means to be a teacher. They adopt the position that the individual teacher has some degree of agency in this process but has limited power to change cultural and socially constructed aspects of this knowledge.

Dillon (2000) also points to the paramount importance of the personal ownership and encouragement in teacher development described in Bell and Gilbert's model but stresses the additional consideration of the role played by others involved in the micropolitics of school science settings. He sees teacher development as a management responsibility. Building on research studies in English settings, he states that the research in teacher development has provided very useful information to date, "but what we lack is a critical understanding and a theory of the management of teacher development" (p. 94). He asserts that teacher development needs support from those who manage teachers and points out that if acts of teaching are complex, management of teaching is even more so. Those who manage teacher development receive little training and suffer from a lack of time to devote to the ethical dilemmas that face them. Dillon provides a list of encouragement factors for successful teacher development in science that are similar to others presented in this chapter, but he adds additional insights, such as the need for sources of new ideas for teachers, an evidence base of successful teaching strategies based on models of learning, and encouragement from managers:

- Initial disturbance or dissatisfaction
- Time to reflect on existing strategies
- An evidence base of successful teaching strategies based on models of learning
- A source of new ideas
- An opportunity to work with colleagues
- Coaching and mutual support
- Appraisal
- Encouragement from managers
- A feeling of personal growth
- A sense of ownership of innovation (p. 95)

In addressing important considerations about the development of self as teacher, we found Maxine Greene's extensive body of work to be a very valuable resource (1987). She argues that development of the self is related to teacher development:

> Teachers are too deeply uncertain when it comes to risking in the name of full human connection because they are uncertain as to what it signifies to realize an ideal of the self. There is a glossing over of the fundamental uncertainties and inequities; there is a dependence on received knowledge; there is a felt need if not to be properly joyful, at least to be or seem to be in proper agreement. For many teachers the problem is one of the divided consciousness; they opt for efficiency and effectiveness because that is what is asked of them; but what they opt for is frequently at odds with what they value and what they believe." (p. 180)

Through her thoughtful articulation of the concerns of teachers, she helps articulate the notion that in their development work, teachers must *think* for themselves, *think* in concert with other professionals and with their students, about the meaning of their work and about how it might be changed in response to those whose lives it impacts while being informed by the discipline being taught. She argues that we must recognize teachers as the authors of their own lives and suggests that teachers must keep questions about their practice open, preferably in association with others so that they may effectively think about what they are doing.

Concluding Remarks

The ultimate goal of this work is to develop a foundation to guide the creation of experiences that will improve science teaching and learning in school settings. Efforts to craft a philsophy for this work has meant expanding ideas about the nature of science, about the ways learners learn it best and about the environments we create for learning science. It also means giving high status to teachers' efforts to link their strong experiential knowledge of classrooms and students and to create new ways to help teachers make meaningful connections to new thinking about science learning, content knowledge, and educational theory. The research summaries and trends that we have reviewed and discussed in this chapter help articulate a foundational position for creating professional development opportunities that we want teachers to see as reflecting authentic professional growth needs and experiences that they can embrace with enthusiasm. The literature we have reviewed, much of it the most current in the field, suggests several useful themes to guide professional development. It points to the inadequacy of outdated models of teaching and learning and suggests new ideas about how students and teachers learn best. Learning anything new is a process of constructing knowledge by integrating new understanding with prior knowledge and beliefs. Elementary teachers are faced with learning about new content knowledge in science and new strategies for helping learners, and they must address new perspectives about the realities of the culture of science learning. Professional development opportunities must address these needs but must also take into account the need for time and the development of structures that allow for new social and personal constructions about what it means to be a teacher. Our experience and findings show that these structures must also take into account the vital need for personal meaning and development of teachers in this process. Mandated curriculum change and transformation is enhanced and new insights into educational theory and practice are facilitated when the kinds of problems teachers consider meaningful are addressed. Teachers must see the value of applying new ideas and approaches to practice. They must feel supported in this effort and be given extended periods of time and opportunities to talk and reflect with peers, colleagues, and others about their efforts to make change and learn new, innovative practices. Our work reemphasizes the contributions of university study and personnel in research collaboration with those involved in staff development opportunities.

With the resources presented here we are building a philosophy that seeks to encourage new, innovative collaborations with teachers to allow opportunities for them to step back from their work; to analyze, discuss, and evaluate new ideas with others in their field; and to reflect on their own practices. We want to create environments that help teachers adopt innovative practices that blend with their own understandings about their work. We believe that this approach and these resources will help guide professional development experiences for teachers that will help them develop greater confidence in their abilities to articulate their own extensive knowledge about what is needed to improve science teaching in schools.

Acknowledgment

This chapter was developed with research resources provided by a grant from the Social Sciences and Humanities Research Council of Canada. This generous support is acknowledged with gratitude.

References

Aikenhead, G. (2000). Renegotiating the culture of school science. In R. Millar, J. Leach, & J. Osborne (Eds.), *Improving science education: The contribution of research* (pp. 245–265). Philadelphia: Open University Press.

American Association for the Advancement of Science. (1993). *Benchmarks for science literacy*. New York: Oxford University Press.

Bell, B. (1998). Teacher development in science. In B. J. Fraser, & K. G. Tobin (Eds.), *International Handbook of Science Education, Part Two* (pp. 667–680). Boston: Kluwer Academic Publishers.

Bell, B., & Gilbert, J. (1996). *Teacher development: A model from science education*. Washington, DC: Falmer Press.

Bencze, L., & Hodson, D. (1999). Changing practice by changing practice: Toward a more authentic science and science curriculum development. *Journal of Research in Science Teaching, 36*(5), 521–539.

Bentley, D., & Watts, M. (Eds.). (1989). *Learning and teaching in school science: Practical alternatives*. Phildelphia: Open University Press.

Brooks, J.G., & Brooks, M.G. (1999). *In search of understanding: The case for constructivist classrooms*. Alexandria, VA: Association for Supervision and Curriculum Development.

Cobern, W.W. (1998) (Ed). *Socio-cultural perspectives on science education: An international dialogue*. Boston: Kluwer Academic Publishers.

Cunningham, D.J. (1984). *What every teacher should know about semiotics*. Urbana, IL: ERIC Clearinghouse on Reading and Communication, National Institute of Education. ERIC Document ED 250 282.

Cunningham, D.J. (1987). Ooutline of an education semiotic. *American Journal of Semiotics, 5*(2), 195–201.

Darling-Hammond, L., & Sykes, G. (Eds.). 1999. *Teaching as the learning profession: Handbook of policy and practice*. San Francisco: Jossey-Bass.

Danesi, M. (1994). *Messages and meanings: An introduction to semiotics*. Toronto: Canadian Scholars Press.

Dillon, J. (2000). Managing science teachers' development. In R. Millar, J. Leach, & J. Osborne, (Eds.), *Improving science education: The contribution of research* (pp. 94–109). Philadelphia: Open University Press.

Duckworth, E., and the Experienced Teachers Group. (1997). *Teacher to teacher: Learning from each other*. New York: Teachers College Press.

Duschl, R. (2000). Making the nature of science explicit. In R. Millar, J. Leach, & J. Osborne (Eds.), *Improving science education: The contribution of research*. (pp. 187–206). Philadelphia: Open University Press.

Greene, M. (1987). Teaching as project: Choice, perspective and the public space. In F. Bolin & F. McConnell (Eds.), *Teacher renewal: Professional issues, personal choices* (pp. 178–189). New York: Teachers College Press.

Greene, M. (1995). *Releasing the imagination: Essays on education, the arts and social change*. San Francisco: Jossey-Bass.

Groisman, A., Shapiro, B., & Willinsky, J. (1991). The potential of semiotics to inform the understanding of events in science education. *International Journal of Science Education, 13*(3), 217–226.

Hargreaves, A. (1995). Beyond collaboration: Critical teacher development in the post-modern age. In J. Smyth, (Ed), *Critical discourses on teacher development* 149–180 Toronto: OISE Press.

Hawley, W.D., & Valli, L. (1999). The essentials of effective professional development: A new consensus. In L. Darling-Hammond, & G. Sykes (Eds.), *Teaching as the learning profession: Handbook of policy and practice* (pp. 127–151). San Francisco: Jossey-Bass.

Kennedy, M.M. (1999). Form and substance in mathematics and science professional development. *National Institute for Science Education (NISE) Brief, 3*(2), 1–6.

Kuhn, T. (1962). *The structure of scientific revolutions*. Chicago: University of Chicago Press.

Last, S. (2001). *Creating a generative curriculum for science teacher development.* Unpublished manuscript.

Lemke, J.L. (1984). *Semiotics and education.* Toronto: Victoria University.

Lemke, J.L. (1990). *Talking science: Language, learning and values.* Norwood, NJ: Ablex Publishing.

Loucks-Horsley, S., Hewson, P.W., Love, N., & Stiles, K. (1998). *Designing professional development for teachers of science and mathematics.* Thousand Oaks, CA: Corwin.

Marx, R.W., Freeman, G., Krajcik J.S. & Blumenfeld, P. (1998). Professional development of science teachers. In B.J. Fraser & K.G. Tobin (Eds.), *International Handbook of Science Education, Part Two* (pp. 667–680). Boston: Kluwer Academic Publishers.

Mundry, S., & Loucks-Horsely, S. (1999). Designing professional development for science and mathematics teachers: Decision points and dilemmas. *National Institute for Science Education (NISE) Brief, 3*(1), 1–7.

National Research Council. (1996). *National science education standards.* Washington, DC: National Academy Press.

National Science Resources Center. (1997). *Science for all children.* Washington, DC: National Academy Press.

Novak, J., & Gowin, B. (1985). *Learning how to learn.* New York: Cambridge University Press.

Polman, J. L. (2000). *Designing project-based science: Connecting learners through guided inquiry.* New York: Teachers College Press.

Richardson, V. (1990). Significant and worthwhile change in teaching practice. *Educational Researcher, 19*(7), 10–18.

Rosebery, A.S., & Puttick, G.M. (1992). *Teacher professional development as situated inquiry: A case study in science education.* Newton, MA: Center for the Development of Teaching Paper Service, Education Development Center. ERIC Document ED 424 084.

Shapiro, B.L. (1994). *What children bring to light: A constructivist perspective on children's learning in science.* New York: Teachers College Press.

Shapiro, B.L. (1996). A case study of change in elementary student teacher thinking during an independent investigation in science: Turning to the "face of science that does not yet know." *Science Education,* 80(5), 535–560.

Shapiro, B.L. (1998). Reading the furniture: The semiotic interpretation of science learning environments. In K. Tobin, & B. Fraser (Eds.), *International Handbook of Science Education* (pp. 600–621). Dordecht, The Netherlands: Kluwer Press.

Shapiro, B.L. (2000). Awakening to objects of meaning in science classrooms. In T. Koballa & D. Tippins, *The promise and dilemmas of teach-*

ing middle and secondary science: A classroom case handbook, (pp. 97–109). Upper Saddle River, NJ: Prentice Hall.

Shapiro, B.L. & Kirby, D. (1998). An approach to consider the messages of science learning culture. *Journal of Science Teacher Education, 9*(3), 221–240.

Shapiro, B.L., Richards, L., Ross, N., & Kendal-Knitter, K. (1999). Time and the environments of schooling. *Learning Environments Research—An International Journal, 2*, 1–19.

Smithsonian Institution Astrophysical Observatory. (2000). *Looking at learning . . . Again.* Cambridge, MA: Harvard-Smithsonian Center for Astrophysics.

Stein, M.K., Smith, M.S., & Silver, E.A. (1999). The development of professional developers: Learning to assist teachers in new settings in new ways. *Harvard Educational Review, 69*(3), 237–268.

Tobin, K., Tippins, D.J., & Gallard, A.J. (1994). Research on instructional strategies for teaching science. In D.L. Gabel (Ed.), *Handbook of Research on Science Teaching and Learning,* (pp. 45–93). New York: Macmillan.

Tobin, K., Roth, W-M., & Zimmerman, A. (2001). Learning to teach in urban schools. *Journal of Research in Science Teaching, 38*, 941–964.

Von Glasersfeld, E. (1988, July–August). *Environment and communication.* Paper presented at the Sixth Meeting of the International Committee of Mathematics Education, Budapest.

Vygotsky, L.S. (1978) *Mind in society.* Cambridge: Harvard University Press.

Chapter 2

Contextualizing Professional Development in Large Multicultural, Multilingual Urban American Communities

PAMELA FRASER-ABDER

In fall 1999 more than 76 million people in the United States were involved in the occupation of education; this included 68.1 million students and 4 million elementary, secondary school, and college faculty, with representation within the classroom from every country in the world. In a nation with a population of about 271 million, more than one out of every four persons participates in formal education. Clearly, from the large number of participants, the many years that people spend in school, and the hundreds of billions of dollars expended by schools, education figures prominently in the life of the nation. Our nation's schools will need to hire 2.2 million teachers over the next decade, and these teachers will need to be well prepared to teach all students to the highest standards in urban, rural, and suburban settings.

Students ordinarily spend from 6 to 8 years in the elementary grades, preceded by 1 to 3 years in nursery school and kindergarten. The elementary school program is followed by 4- to 6-year secondary school program. The elementary program is frequently followed by a middle school or junior high school program, which generally lasts 2 or 3 years. Students then may finish their compulsory schooling at the secondary or high school level, which may last from 3 to 6 years, depending on the structure within their school district. Students normally complete the entire program through grade 12 by age 17, 18, or 19.

High school graduates who decide to continue their education may enter a technical or vocational institution, a 2-year college, or a 4-year college or university. A 2-year college normally offers the first 2 years of a standard 4-year curriculum and a selection of terminal vocational programs. Academic courses completed at a 2-year college are usually transferable for credit at a 4-year college or university. A technical or vocational institution offers postsecondary technical training leading to a specific career. Community organizations, libraries, churches, and businesses offer other types of educational opportunities for adults.

An associate's degree requires the equivalent of at least 2 years of full-time college-level work, and a bachelor's degree normally can be earned in 4 years. At least 1 year beyond the bachelor's is necessary for a master's degree, while a doctoral degree usually requires a minimum of 3 to 4 years beyond the bachelor's.

Professional schools differ widely in admissions requirements and in program length. Medical students, for example, generally complete a 4-year program of premedical studies at a college or university before they can enter the 4-year program at a medical school. Law programs normally require 3 years of coursework beyond the bachelor's degree level. (See Figure 2.1.)

Generally, students take science courses from kindergarten through grade 12, with general science in K–6 and one year of biology, earth science, chemistry, physics, life sciences, or physical science in grades 7–12. The more academic schools allow their students to complete Advanced Placement science classes in biology, chemistry, and Physics. The content taught is determined at the state level and is largely influenced by national standards.

TIMSS and Achievement in Science

During the 1990s international comparisons of student achievement became the venue for discussion about the quality of American public schools. Results from The Third International Mathematics and Science Study-Repeat (TIMSS-R) showed that U.S. eighth graders performed higher than 18 countries, similarly to five countries, and lower than 14 countries in science. Countries with the highest performance in science were China, Singapore, Hungary, Japan, and the Republic of Korea. Students from homes with high levels of educational resources were among the top achievers, with the lowest achievers coming from four urban districts that also had the lowest percentage of students with high levels of home educational resources. These results support the extensive research showing that students in urban districts also often attend schools with fewer resources than nonurban districts, including a less challenging curriculum and less qualified teachers.

Following the release of TIMSS results, the questions most frequently posed were: Why did American students fare poorly, and what could be done to help students in this country achieve "world-class standards"? Major findings indicated that in the United States, students were more likely than their international counterparts to be taught science by teachers with degrees in education or "other" than by teachers with a bachelor's or master's degree in science. Only 27 percent of the eighth-grade teachers in the benchmarking entities said they felt "very well prepared" to teach science. TIMSS results remind us that improving students' opportunities to learn requires examining

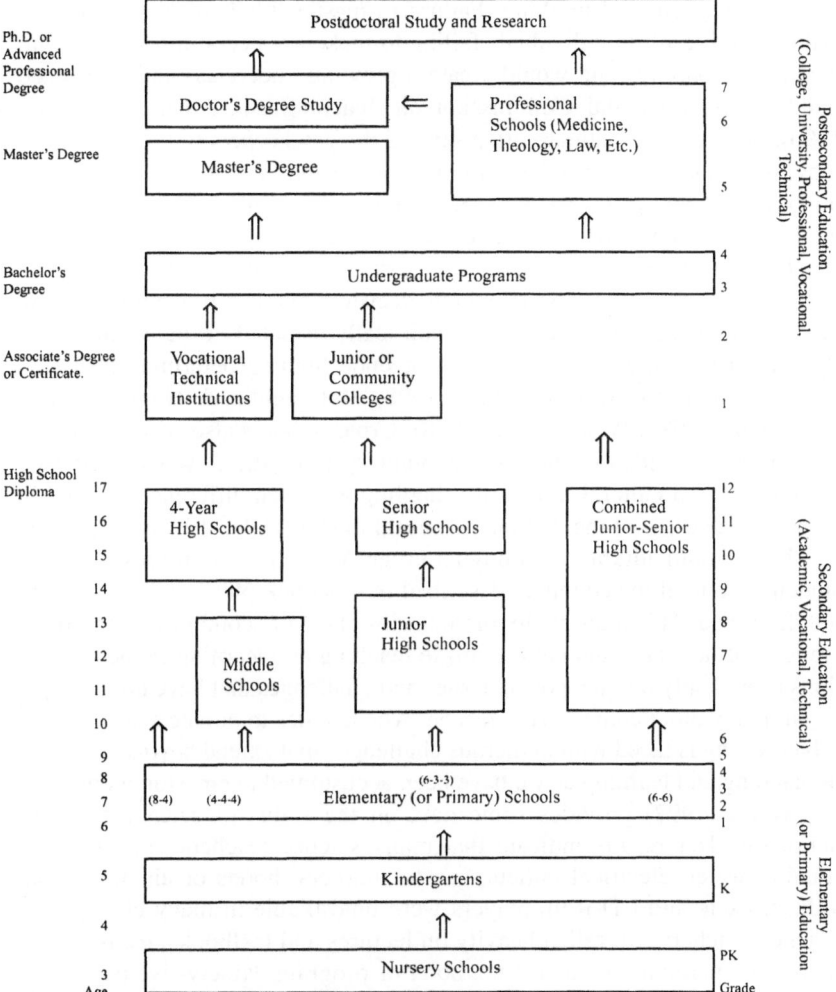

Figure 2.1. The structure of education in the United States.
Note: Adult education programs, while not separately delineated here, may provide instruction at the elementary, secondary, or higher education level. Chart reflects typical patterns of progression rather than all possible variations. *Source:* U.S. Department of Education, National Center for Education Statistics.

every aspect of the educational system, including teacher quality; hence the need to reform and expand professional development of teachers is great.

For at-risk students in urban schools, mastering "world-class standards" was assumed to be particularly critical because traditional blue-collar work was being replaced by jobs that required much more advanced subject area content, research, and thinking. Failure to master this knowledge and new skills would result in individuals failing to make a successful transition into the "new economy" and would create a growing underclass and threatened middle class (National Commission on Teaching and America's Future, 1996). Science and math are often seen as the gatekeepers to this new economy; subsequently, the need to improve science and math teaching and learning has led the way in the school reform movement with particular emphasis on the urban schools where the needs are greater.

Urban schools serve a large, highly diverse population (Council of the Great City Schools, 1995) in which decision making is centralized and invested in a bureaucracy that is politically isolated from communities (Rogers, 1968) and chronic patterns of underfunding determine decisions about teaching and learning (Krei, 1998; McLaughlin & Hopfengardner, 1998; Mirel, 1993; Wong & Lee, 1998). Urban schools also serve high concentrations of students who are "involuntary minorities," whose linguistic capabilities and cultural model of schooling are often different from and in conflict with those of the dominant cultural model (Ogbu, 1995a, b).

Large multicultural, multilingual urban American classrooms comprise students from all the countries described in this book plus hundreds that are not described. This multiplicity of countries of origin, combined with proliferations of language and culture, make teaching and learning in these situations remarkably unique, posing issues and challenges that have not yet been faced in any other country. Any teacher who decides to pursue a career in an urban setting is faced with numerous challenges that extend beyond the aegis of teaching and learning as we have been accustomed to envision them.

Weiss (1997) provides some data on the status of science teaching nationally. Her results indicate that many science teachers indicated that running water, electrical outlets, gas for burners, hoods or air hoses, and videocassette and CD-Rom players were unavailable in many classrooms. Science teachers still relied heavily on lectures and textbooks, using one or more commercially published textbook or program. Ninety-five percent of middle and high school science classes used commercially published textbooks or programs. At the elementary level, 75 percent of science classes used published textbooks or programs, down from 86 percent in 1986. The largest proportion of class time was devoted to lecture and discussion (38%), followed by hands-on or laboratory work (23%), individual seat work (19%), and nonlaboratory small-group work (10%), with the remaining 10 percent of time spent on non-instructional activities. Sixty percent of high school science and math classes listen and take notes during presentation by the teacher on a daily basis; 94 percent do so at least once a week. More than 60 percent of high school science classes never take field trips, 54 percent never use computers, and 43 percent never work in class on science projects that last at least a week. The most heavily emphasized objectives in science classes were learning basic science concepts (heavily emphasized in 83% of science classes overall), increasing awareness of the importance of science in daily life (77%), and developing problem-solving and inquiry skills (74%). Approximately 20 percent of science classes in each grade range put heavy emphasis on preparing students for standardized tests. Teachers in classes with high proportions of minority female students are more likely than others to emphasize preparing students for standardized tests and less likely to aim toward preparing students for further study in science and math.

Many science teachers were underprepared, underpaid, and overworked. One quarter of newly hired American teachers lacked the qualifications for their jobs. More than 12 percent of new hires entered the classroom without any formal training; another 14 percent arrived without fully meeting state standards. More than 40 states allowed districts to hire teachers who had not met basic requirements. In recent years, more than 50,000 people who lack the training for their jobs have entered teaching annually on emergency or substandard licenses. Nearly one quarter of all secondary teachers do not have a minor in their main teaching field. Among teachers who teach a second subject, 36 percent are unlicensed in the field and 50 percent lack a minor.

Following the release of the TIMSS report we have seen many attempts to solve the problems highlighted by Weiss (1997). National science and mathematics standards have been introduced, new teacher certification and accreditation guidelines have been instated, and more money has been allocated to professional development of teachers. The recently released 2000 National survey of 6000 teachers by Horizon Research Inc. (2002) (http://2000survey.horizon–research.com/reports/paemst.php) indicate that more than 80% of science lessons in grades K-12 include discussion and lecture and that most teachers perceive a need for professional development particularly in the use of technology for instruction.

This chapter is limited in its scope drawing primarily from a unique small setting in one of the largest urban populations in the United States. New York state has recently joined the nation in its search for solving the dilemma faced by students and teachers in the school system. The Regents, the state governing board on education, recently released their plan (1998) for revamping teacher education and have provided us with some startling information about education in New York. New York state is known for outstanding education, (Students wrote 11 percent of Advanced Placement exams, comprised 9 percent of students commended by the National Merit Scholarship Program, were 40 percent of the winners in the Westinghouse Science (now Intel) talent search, and scored 10 points higher than the national average on the combined subject tests of the SAT.) but at the same time, education results in some areas of the state are shockingly poor.

Status of Science Teaching in New York City in 1997

At the beginning of the 1997–1998 school year, 99 schools had been placed under registration review (SURR) for continuing poor student performance; 92 of these schools were in New York City. In 1994–1995 the graduation rate statewide was 66.1 percent, 46.7 percent in New York City and 81.1 percent in the rest of the state.

In 1996, the New York State Board of Regents initiated a sweeping reform of the way they recruit, prepare, certify, and continue to educate teachers. The Regents' strategic plan challenges the entire educational system to meet their primary goal of having all students meet high standards for academic performance and demonstrate the knowledge and skills required by a dynamic world. To accomplish this goal the State Education Department commissioned the Regents' Task Force on Teaching to assess the problem areas and propose a strategic plan to guide the education reform (Teachers for Tomorrow, 1997). According to the task force major problems existed within the New York State and New York City education system. On average students in New York City scored 26 percent lower on third-grade and 23 percent lower on sixth-grade state reading tests than the rest of the state did.

There are 202,000 teachers in New York state; approximately 10,000 of them are uncertified and teaching with temporary licenses. More than 9,000 of the uncertified teachers are teaching in New York City, while there are only 629 in the rest of the state. Of the 75,000 teachers in New York City it is projected that 15,000 teachers will retire by 2003, and the city expects an increase in enrollment size that will demand a need of 15,000 new teachers, bringing the total to 30,000 teachers over the next five years. Certification, a license to teach in New York State public schools, is now virtually permanent and inalterable. Teachers can retain certification throughout their career without having to demonstrate continuing competence or professional growth. Hence there was no need for continued professional development beyond certification.

Further analysis of the problems revealed critical competency gaps in New York State:

1. New York State does not attract and keep enough of the best teachers where they are needed most.
2. Not enough teachers leave college prepared to ensure that New York's students reach higher standards.
3. Not enough teachers maintain the knowledge and skills needed to teach to high standards throughout their careers.
4. Many school environments actively work against effective teaching and learning. (Regents Task Force Report, 1998)

Another problem that emerged as a policy concern late in the decade, one that dominated headlines of city newspapers, was the shortage of qualified teachers, especially in schools that served the highest concentrations of students from multicultural, multilingual communities and families in poverty. In New York City, the courts ordered a freeze on hiring of certified teachers by more successful schools so that underperforming schools would have an opportunity to hire them first (Hartocollis, 2000). A large exodus of newly certified teachers, particularly in science and mathematics, is expected in 2001 and beyond; these teachers will go to suburban schools, which offer superior salaries and working conditions.

Cultural, Equity, and Contextual Considerations

In the past two decades 90 percent of the nation's newcomers have come from Latin America, Asia, Africa, and the Caribbean. Most, therefore, are members of groups with whom the public schools historically have been neither comfortable nor successful (Seller & Weis, 1998). A report issued by the Council of the Great City Schools in 1995 noted that about half of the nation's new immigrants in the 1980s had gone to Los Angeles, Chicago, Washington, DC, New York, Houston, and San Francisco; over 90 percent of the new arrivals to this country had gone to metropolitan areas (Council of the Great City Schools, 1995). By far the largest group of immigrants to U.S. schools in the decade was Spanish-speaking, many from Mexico. This massive number of newly arrived students present the urban school systems serving them with previously unknown and unheard of organizational administrative challenges. Teachers require special skills and knowledge when working with newly arrived immigrants, especially in communities and school settings that have not adapted to their presence (Valdes, 1998). Scholarship is surprisingly minimal about how school systems, teachers, and programs of teacher preparation might respond to challenges beyond language acquisition to deal with the persistent life situations of the newly arrived immigrant child (Igoa, 1995).

Nationally, the proportion of science and math teachers who were themselves members of minority groups is low—11 percent in the elementary and middle grades and 7 percent at the high school level; roughly 30 percent of students belong to minority groups. In most states, the number of minority teachers equal one-third or less the number of minority students. The southeastern states, California, and Hawaii now have the highest proportions of science teachers from minority populations. In New York City, 83 percent of the students are minorities compared to 34 percent of the teachers; the rest of the state has a 43 percent student minority population, and 15 percent of the teachers are minorities.

Teacher education students continue to admit, as they have earlier, that they want to teach students in familiar settings and who are like themselves and that they often are uncomfortable with personal contact with parents from ethnic and language minorities (Gomez, 1996; Zeichner & Baker, 1995). Teacher candidates are "parochial," while their classroom population is becoming increasingly cosmopolitan (Zimpher & Ashburn, 1992). Countless articles and papers have detailed the difficulty of changing teachers' attitudes and practice as they worked with students who were not white, middle class, and monolingual, as most teacher candidates are (Bixby, 1997; Vacc, 1998; Winitzky & Barlow, 1998).

Urban teachers often work detached from the community and family resources that would help them to understand their student's lives, needs, and interests, a problem compounded by procedures and regulations designed to make education impersonal and anonymous. Moreover, schooling's white, middle-class cultural norms and the inequalities of power relationships between teachers and parents are veiled and reinforced by bureaucratic procedures and norms, subverting authentically caring relationships (Hamovitch, 1996).

Life for students is a series of problems, a series of decisions. Students bring to the science classroom their own problems, many of which are influenced by the system, their home environment, peers, and teachers, and which often result in students leaving school without a basic understanding of science. Teachers need to be aware of the variety of the problems their students confront on a daily basis, problems that to many teachers often appear to be insurmountable.

Culturally relevant science is critical to student participation and achievement. Science teachers of many years are now being expected to acknowledge and accept students of cultures and perspectives different from their own and to include in their curriculum material that may be unfamiliar and often uncomfortable. If teachers are to make science culturally relevant to students they must be sensitive to the beliefs and learning as conceptualized within the context of the lives of the students in their classes. Many teachers belong to cultures that are different from the culture of their students, do not understand the issues that are prevalent in the culture of their students, and often demonstrate a lack of sensitivity to these issues. On many occasions their preconceived perceptions of certain cultures leads to stereotyping of expected behaviors and norms that is reflected in their teaching.

Urban teachers in the 2002 are less likely to be working with white students and more likely to be teaching immigrants, and when teaching African American and Hispanic students might well be working in schools that are not racially but culturally diverse. The most salient aspect of urban teaching then is that urban teachers must be able to accommodate the greatest diversity of student needs under conditions that continually subvert their efforts to personalize and individualize education. Student diversity is increasing, but teacher diversity and preparation for teaching a diverse student body is not keeping pace. Certification regulation requires teachers to have only minimal preparation for teaching students of diverse characteristics and backgrounds; often one course is provided to cover the requirement or the topic is superficially woven through a few courses.

It is within this context that teachers are prepared to teach science. How can teacher educators empower teachers to cope with the continuously evolving diversity in their classes while increasing interest and achievement in science among their students? What are some of the significant areas that need to be addressed to foster changes in the way science is currently being taught in the schools? How can teachers make science accessible and inviting to their students? What is the appropriate model for providing professional development for science teachers in the urban environment? These are some of the issues science educators have struggled with as they attempt to reenvision professional development through the lens of culture, equity, and contextual considerations.

Empowering Teachers to Cope with the Continuously Evolving Diversity in Their Classes while Increasing Interest and Achievement in Science

"Every student has a right to be taught by caring and competent professionals" (Witty, 1998, p. 11). Everyone, from the president of the United States to policy makers and parents, has echoed these words. Given this national cry for competent professionals, how can we best prepare future teachers for work, particularly in urban science classrooms where the need is greatest? Witty suggests that we

- Include extensive experiences with urban children and parents throughout the training program.
- Promote the understanding that teacher success is tied to student success—teacher education graduates must understand that they are not good teachers if their students are not achieving.
- Provide continuous experiences for students in self-assessment and reflection.

Preparing teachers to become successful science teachers is an ambitious undertaking. Teachers are being prepared for a world in which they have not yet lived and can only presuppose the conditions that will exist. They need to understand the complexities that make every teaching situation unpredictable and unique. They need to be exposed to the multiplicity of theories of teaching and yet acknowledge that good theory does not always guarantee good learning. They must develop the ability to observe and reflect on their own practice and to be sensitive to the needs and feelings of all their students.

They must experience "being different" before they can identify with the feelings of their culturally diverse students (Fraser-Abder, 2001). Most of our teachers come from a culture that is different from the culture to which their students belong. They are often unaware of the cultural norms that exist in these diverse cultures, which oftentimes interferes with the teaching and learning of science in the American culture. They come to the classroom with often unconscious stereotypes for the students they encounter, and their teaching and expectations often reflect these beliefs.

While issues of how best to prepare teachers for an increasingly diverse student population, including students historically underserved by public education, are of particular concern to those who prepare prospective urban teachers, the issue of preparing teachers to work successfully with students different from themselves should be a concern of all programs of teacher preparation. Preparing teachers to teach students whose values and life experiences are quite different from their own, to be knowledgeable about racism and its effects on educational achievement, should not be the exclusive responsibility of urban teacher preparation. All programs of teacher preparation should embody these goals.

As indicated by Fraser-Abder (2001), teachers in culturally diverse classrooms should:

- Learn as much about and become as sensitive to and aware of racial, ethnic, cultural, and gender groups other than their own as they can.
- Never make assumptions about an individual based on their perception of that individual's race, ethnicity, culture, or gender.
- Avoid stereotyping.
- Get to know each student as an individual: *Walk in the footsteps of all their students.*

A professional development model for helping teachers better understand the needs of their students and subsequently use appropriate pedagogical strategies for meeting these needs and allowing science to be more accessible to such students has been described by Fraser-Abder (2001). The overarching goal for this professional development model was *To empower teachers to identify, and find solutions to, the complex educational problems they face in their science classrooms and to create innovative, effective, and culturally relevant science curricula and pedagogy.* The specific goals of the program were to:

1. Provide science teachers with the tools they need to teach science/math to the continuously evolving diverse population in their classrooms.
2. Improve the quality of science teaching and increase interest, achievement, and learning for all students.
3. Develop and teach culturally relevant science curricula.

This model has been successfully used with over 300 graduate students in the science teacher education program.

As science teacher educators begin to reexamine and revise the content of professional development programs, a cultural contextual framework for overcoming hurdles within the urban science classroom must form the core of resultant programs, and participants must be provided with the skills to:

1. Begin to assess their own conscious and subconscious biases about people who are different from themselves in race, ethnicity, culture, gender, or socioeconomic status.
2. Plan their science curriculum within a multicultural framework while making their classroom a safe and secure haven for all students.
3. Infuse multicultural instructional materials and strategies in their science teaching.
4. Foster collaboration and cooperation among students, parents, teachers, and administrators.

Significant Areas That Need to Be Addressed to Foster Changes in the Teaching and Learning of Science

Comparing school organization in four high schools in a metropolitan area, Metz (1997) found that the nonacademic tasks required of all schools, such as keeping students safe, were far more time-consuming at the urban schools in poor neighborhoods. She concluded that money was used for custodial concerns, aides and administrators who kept corridors clear and maintained security in the urban schools, whereas in the suburban schools serving wealthier students resources were used for professional development, like attendance at conferences.

Winfield and Manning (1992) identified many common practices in urban schools that undercut teacher initiative and sense of efficacy, including the existence of separate, uncoordinated instructional programs that encouraged a custodial attitude toward children and the belief that teachers were simply referral agents. They provide evidence that changes in supporting structures are essential to alter existing norms and beliefs of both teachers and students. Others studies demonstrated that the urban context, the community, and the school system greatly influence and even configure learning, instruction, and curriculum in urban classrooms.

However, teachers will not change unless they firmly believe that they must do so if they are to be successful teachers. As students attempt to make connections between their life at home, their life at school, and their life as they make the journey between school and home, they are, in fact, leading three separate lives, with the referees in each life not being cogniscent of the other lives, and therefore not in a position to help students make the transition from one life to another or to make relevant connections between these lives. Students see their status in these lives as being very disconnected. To them there are no relevant relationships between these lives. It is only when the referees in each life (parents, teachers, and the community) are cogniscent of the other lives that they will be in a position to help students make the transition from one life to another, to make relevant connections between these lives and to reap the benefits of learning science. It is only then that teachers will be successful.

Making Science Accessible and Inviting to All Students

Reforming teacher preparation to educate teachers who are more knowledgeable and skilled in pedagogy and subject matter was essential for reforms of K–12 education designed to hold students accountable for "world-class" standards.

Teachers are increasingly being faced with classroom situations for which they have not been prepared and which they themselves have never previously encountered (Loving & Marshall, 1997). It should be noted that although data have been presented for New York City, the same holds true for most urban settings in which teachers are required to engage in multicultural science education defined by Atwater (1996) as a field of inquiry with constructs, methodologies, and processes aimed at providing equitable opportunities for all students to learn quality science. In this context it is imperative for professional development to include strategies for understanding how class, culture, disability, ethnicity, gender, language, and power influence the learning and teaching of science.

Today we need to develop science teachers who can function with technology, who can deal with social and family problems, who can respond to the learning styles of diverse populations, and who have the skills and content knowledge to teach to high standards. The diversity present in urban school settings requires teachers to renegotiate their position on curricular and pedagogical issues. Good pedagogical practices should address the first four deficiency needs: physiological, safety, esteem, and emotional (Maslow, 1971). It is not a revolutionary concept to think that teachers must be in touch with all aspects of their students' lives. However, the design and subsequent implementation of this concept is a novel idea in many classrooms around the country. Schools of education and other professional development organizations must provide teachers with training that makes them more aware of the students' deficiency and growth needs, as expressed by Maslow; additionally, teachers need opportunities to research and to evaluate how these needs transcend the classroom.

Science will be accessible and inviting to students when

- Teachers are involved in innovative ways of teaching science, with the full support of the community and the parents.
- Universities are actively involved in research in the classrooms.
- Teachers are involved in action research and in experimenting with effective strategies for science teaching and learning.
- Teachers write and obtain grant funding to implement effective programs.
- Teachers use the latest developments in technology to enhance teaching.
- Teachers foster an environment conducive to achievement in science.

However, this status quo does not develop overnight; it takes time, commitment, and hard work by administrators, teachers, parents, the community, and students.

Appropriate Models for Providing Professional Development for Science Teachers in the Urban Environment

There is no one model that can be replicated in all urban science classrooms. Professional development must expose teachers to strategies that have worked, and it must give them opportunities through action research to develop their own strategies. Austin and Fraser-Abder (1995) suggest some philosophical underpinnings for science teacher professional development programs in large urban multicultural, multilingual settings and some strategies for involving teachers in the successful teaching and learning of science. They suggest that the underlying philosophy of any emerging professional development program must embrace the belief that:

1. For teacher education to be *successful* it must be informed by classroom practice in the schools for which teachers are being prepared.
2. For teacher education to have a *long-term effect* teachers must experience success in the practical application of the theoretical knowledge they acquire in their university courses; they must experience their students succeeding in and enjoying science. They must see the relevant applicability and transferability of the content and pedagogy taught by the university professor to their own classroom situation.
3. For teacher education to be *effective*, instruction should be structured in a supportive peer and classroom environment in which courses are taught using strategies that teachers perceive as being appropriate for modeling in their own classrooms.

4.   For teacher education to result in *appropriate* learning, it must inculcate in teachers an in-depth knowledge and appreciation of the role of gender, culture, and psychosocial factors in the teaching and learning paradigm.

Teacher educators need to develop teacher enhancement and professional development activities that improve teachers' abilities to help all students achieve high performance standards and which are governed by the following assumptions:

1.   Every child can learn and is capable of learning, if provided with the right teaching and learning environment.
2.   No two children are alike.
3.   Strategies that work with one child might not work with another.
4.   Students' background and experiences should be considered when teaching basic academic concepts.
5.   Community members from various ethnic groups can assist teachers in confronting issues of ethnic differences and similarities. (Fraser-Abder, 2001).

Creation of professional development schools (PDS) was the strategy most often advocated to improve the preparation of urban teachers. Generally the PDS involved using a select group of school sites to become the nexus of collaboration between schools and universities, providing teacher candidates with practical experience in school settings in which they could see exemplary teaching modeled, while providing schools with assistance from university faculty in areas such as curriculum development. Research, however, did not demonstrate that the PDS actually improved teacher preparation or academic achievement of urban students. One national survey of participants in professional development schools found that although "many people are making valiant efforts to support professional development schools . . . the effort is time intensive and requires long-term individual commitment. Although positive comments and numerous benefits are being reported, the hoped-for 'reinvented institutions' do not seem to be forming" (Kochan, 1999, p. 187). However, in New York City much sustained effort is underway in PDS with many groups reporting immense success (Walsh, 2001).

Conclusion

Teachers must confront their own cultural realities and explore ways that their interpretation of culture influences their teaching and interactions with students. As classrooms grow more challenging and diverse, teachers will need to be well prepared to teach all students to achieve the highest standards. Schools in high poverty urban areas have a particularly pressing need for greater numbers of well-prepared teachers. Contemporary classrooms and social conditions confront teachers with a range of complex challenges previously unknown in the profession. These challenges mean that high-quality teacher preparation and development is more important than ever. Both incoming and current teachers need to develop a better understanding of the persistent life situations of their students. They need to get to know their students before they can develop teaching and learning strategies for those students. Teachers must understand that they cannot wear the skin of their students; at this point in their lives it is their time to wear that skin and to deal with the pressures involved in so doing (Fraser-Abder, 2001). However, as teachers you can walk in their footsteps and begin to understand what they experience as they wear that skin and be sensitive to these experiences as you develop learning opportunities for each student.

References

Atwater, Mary M. (1996). Social constructivism: Infusion into the multicultural science education research agenda. *Journal of Research in Science Teaching, 33* (8), 821–837.

Austin, & Fraser-Abder, P. (1995). Mentoring mathematics and science preservice teachers for urban bilingual classrooms. *Education and Urban Society, 28* (1), 67–89.

Bixby, J. (1997, March). *School organization as a context for white teachers' talk about race and student achievement.* Paper presented at the annual meeting of the American Educational Research Association, Chicago.

Council of the Great City Schools. (1995). *Immigrant and limited English proficient youth: Issues in the Great City Schools* (ERIC ED390954). Washington, DC: Author.

Fraser-Abder, P. (2001). Preparing science teachers for culturally diverse classrooms. *Journal of Science Teacher Education, 12*(2), 123–131.

Gomez, M.L. (1996). Prospective teachers' perspectives on teaching "other people's children." In K. Zeichner, S. Melnick, & M. L. Gomez (Eds.), *Currents of reform in preservice teacher education* (pp. 109–132). New York: Teachers College Press.

Hamovitch, B. (1996). Socialization without voice: An ideology of hope for at-risk students. *Teachers College Record, 98*(2), 286–306.

Hartocollis, N. (2000, September 8). Schools try to smooth the bumps on day 1. *New York Times,* pp. B1, B8.

Igoa, C. (1995). *The inner world of the immigrant child.* Mahwah, NJ: Lawrence Erlbaum Associates.

Kochan, F.K. (1999). Professional development schools. In D.M. Byrd & D.J. McIntyre (Eds.), *Research on professional development schools. Teacher education yearbook VII* (pp. 173–190). Thousand Oaks, CA: Corwin Press.

Krei, M.S. (1998, April). *Inequalities in teacher allocation: Policy and practice in urban school districts.* Paper presented at the annual meeting of the American Educational Research Association, San Diego.

Loving, C.C., & Marshall, J.E. (1997). Increasing the pool of ethnically diverse science teachers: A mid-project evaluation. *Journal of Science Teacher Education, 8*(3), 205–217.

Maslow, Abraham H. (1971). *The farther reaches of human nature.* New York: Harper.

McLaughlin, M.J., & Hopfengardner, W.S. (1998, April). *Building capacity for reform in urban schools: Lessons learned through special education.* Paper presented at the annual meeting of the American Educational Research Association, San Diego.

Metz, M.H. (1997, April). *Keeping students in, gangs out, scores up, alienation down, and the copy machine in working order: Pressures that make urban schools in poverty different.* Paper presented at the annual meeting of the American Educational Research Association, Chicago.

Mirel, J. (1993). *The rise and fall of an urban school system: Detroit, 1907–81.* Ann Arbor: University of Michigan Press.

National Commission on Teaching and America's Future. (1996). *What matters most: Teaching for America's future.* U.S. Dept. of Educ. Washington, DC.

Ogbu, J.U. (1995a). Cultural problems in minority education: Their interpretations and consequences. Part one: Theoretical background. *The Urban Review, 27,* 189–205.

Ogbu, J.U. (1995b). Problems in minority education: Their interpretations and consequences. Part two: Case studies. *The Urban Review, 27,* 271–297.

Regents Task Force Report. (July 16, 1998). *Teaching to higher standards: New York's commitment.* Albany, NY: NY State Education Department.

Rogers, D. (1968). *110 Livingston Street: Politics and bureaucracy in the New York City Schools.* New York: Random House.

Seller, M., & Weis, L. (1998). Immigrants and education: An introduction to the special issue. *Educational Policy, 12,* 611–614.

Teachers for Tomorrow. Report of the Regents' Task Force on Teaching. Albany, NY NY State Education Department

Vacc, N. N. (1998, April). *Changing teachers' beliefs through professional development.* Paper presented at the annual meeting of the American Educational Research Association, San Diego.

Valdes, G. (1998). The world outside and inside schools: Language and immigrant children. *Educational Researcher, 27*(6), 4–18.

Walsh, M. (2001). *The professional development laboratory.* New York: New York University.

Weiss, Iris R. (1997). The status of science and mathematics teaching in the United States: Comparing teacher views and classroom practice in national standards. *NISE Briefs, 1*(3).

Winfield, L.F., & Manning, J.B. (1992). Changing school culture to accommodate student diversity. In M.E. Dilworth (Ed.), *Diversity in teacher education* (pp. 181–214). San Francisco: Jossey-Bass Publishers.

Winitzky, N., & Barlow, L. (1998, April). *Changing teacher candidates' beliefs about diversity.* Paper presented at the annual meeting of the American Educational Research Association, San Diego.

Witty, E. (1998, November). *Teacher preparation: Framing the issues from a national perspective.* Paper presented at the Challenge 2000: Recruiting and Preparing Teachers for Urban Schools, NYS K–16 Professional Development Network, Brooklyn, New York.

Wong, K., & Lee, J. (1998, April). *Interstate variation in the achievement gap among racial and social groups: Considering the effects of school resources and classroom practices.* Paper presented at the annual meeting of the American Educational Research Association, San Diego.

Zeichner, K., & Baker, B. (1995). Processes for leadership and change: Teacher leadership for urban schools. Overview and framework. In M.J. O'Hair & S.J. Odell (Eds.), *Educating teachers for leadership and change. Teacher education yearbook, vol. III* (pp. 71–76). Thousand Oaks, CA: Corwin Press.

Zimpher, N., & Ashburn, E. A. (1992). Countering parochialism among teacher candidates. In M.E. Dilworth (Ed.), *Diversity in teacher education* (pp. 40–62). San Francisco: Jossey-Bass Publishers.

Internet Reference

http://2000survey.horizon–research.com/reports/paemst.php

Chapter 3

Professional Development in Japan and China

Issues and Challenges

PAMELA FRASER-ABDER AND SAU-LING CHEN

Japan's Education System

The Japanese education system is highly centralized and is administered by the Monbusho, or Ministry of Education. At the regional level each prefecture or major urban district has its own board of education. In some areas, high schools are under the direct authority of these boards, while elementary and junior high schools are managed by local or municipal boards of education.

The school system from kindergarten through university serves about 24 million students, with about 10 percent of these at the university level (Japan Ministry of Education, http://www.mext.go.jp/english/statis/index.htm). About one-third of all students are enrolled in private schools, and the remainder are in the public or national school system. About 95 percent of preschool children are enrolled in some type of preschool program, either kindergarten or daycare, with about 80 percent in private facilities.

Education is compulsory for students between ages 6 and 15 (i.e., for those at elementary and junior high school levels), and tuition is free for students in public or national schools. In these schools parents must provide for materials such as uniforms, math kits, and calligraphy sets. In elementary, junior, and senior high school, overall enrollment is roughly 90 percent in the public/national schools, but there is an increasing tendency toward private enrollment at the higher education levels. Nearly 30 percent of senior high school students are enrolled in private schools, and at the junior college and university levels, nearly two-thirds are enrolled in private institutions.

Additionally, there are special schools at all levels for students with severe disabilities as well as a small number of elite nationally funded 5-year high schools. Enrollment in these schools represents only a very small percentage of the overall enrollment.

Elementary School

The first six years of compulsory school comprise the elementary grades. Elementary schools are generally within walking distance of children's homes. The elementary curriculum is divided into three major areas: regular subjects, moral education, and special activities. There are nine regular subjects: Japanese, social studies, arithmetic, science, life and environmental studies, music, arts and handicrafts, homemaking, and physical education. Special activities, including clubs, schoolwide festivals, competitions, student associations, and student-run activities, are an important part of the overall curriculum, and teachers spend considerable amounts of time organizing and participating in these and similar activities. The official school year for elementary school lasts 35 weeks. On average, there are 35 students per class (Adelman, 1998). Class periods are 45 minutes long with 10-minute breaks between. The number of class periods increases with each grade. The first three grades are scheduled for 850, 910, and 980 periods, respectively. The upper three grades are scheduled for 1,015 periods. The typical school day starts at 8 A.M., and classes end at 3:50 P.M. Homeroom meetings occur at the start and end of each day, and about 2 hours are spent in recess, lunch, and cleaning.

Junior High School

After six years of elementary school, students enter junior high school for 3 years. Classes are organized by subject, and teachers rotate from class to class while students stay in their classrooms. Teachers are organized according to grade, corresponding to class year, as well as by academic subject and by the committees on which they serve. The grade divisions are the most important, giving the students and teachers a strong sense of belonging to the homeroom class and grade.

School periods are 50 minutes long, and the school year is scheduled for 1,050 periods. Special activities take up between 35 and 70 hours in the first year and progressively decrease in the higher grades. About 94 percent of junior high school students attend public schools, and over 96.8 percent continue on to high school (Japan Ministry of Education http://www.kantei.go.jp/foreign/link/link3.html#21). Since public high school education is neither compulsory nor free, the third year of junior high school is concentrated on students' preparation for the high school entrance exam. In the vast majority of junior high schools, extra classes are held before and after school, as well as during the holidays to provide students with extra preparation for the high school entrance exam.

High School

Although education at the high school level is neither compulsory nor free, each prefecture or municipal district maintains publicly funded high schools that offer relatively low-cost education. About 96 percent of junior high school students continue on to high schools. However, about one-third of all high school students attend private institutions. High school students are admitted based upon high school entrance exam scores, although different prefectures place different emphases on grades and test scores.

The vast majority of the public and private high schools are 3-year institutions. In addition, there are "night schools," correspondence courses, and nationally funded 5-year schools, but these constitute less than 5 percent of overall enrollment.

About 75 percent of students are enrolled in academic courses, 25 percent in vocational courses. Few vocational school graduates apply to attend college because the vocational courses do not offer the rigor necessary for the college exam.

High school periods last 50 minutes and are scheduled for 1,190 hours. Clubs and other extracurricular activities also occupy a significant amount of students' time. Extra classes are commonly scheduled for the academic high schools, but not for the vocational schools. Most vocational school students take part-time jobs and enter the workplace upon graduation. Nationally, about 46.2 percent of high school students continue on to university.

Shadow Education

Outside the school system exists a dense network of academic institutions, consisting of home tutors, correspondence schools, *juku,* and exam prep schools. These extra-school forms of education are referred to as *juku* and shadow education because their curriculum tends to shadow the formal curriculum of the public schools.

The two major types of *juku* are individual enrichment *juku* and academic *juku.* Individual enrichment *juku* courses are primarily nonacademic, consisting of such activities as swimming, piano, or calligraphy. These classes are most common among elementary school students.

Academic *juku* can be either:

• Review *juku*—catering to students requiring remedial assistance. Review *juku* are popular with junior and high school students who get help in subjects they find difficult. Many of these *juku* are run out of private homes and use the same texts used in the classroom.
• Advancement *juku*—catering to students preparing for the entrance exams. Junior and high school students preparing for the entrance exams largely attend these.

Students who fail to get into college may also spend up to two years studying at special schools—yobiko—that specifically prepare students for the college exams. The fact that most Japanese businesses would hire some-one based on the college or university from which he or she graduated rather than the grades he or she earned puts a great deal of pressure on students to enter the most prestigious institutions. Some of these students will continue to study and make additional attempts for higher education until they decide to resign from their desperate pursuits (Petterson, 1993).

The enormous amount of emphasis on school entrance exams have led many students to suffer from what is known as the "School Refusal Syn-drome." Students become fearful of attending school and begin cutting classes. Other students suffer from more serious effects, such as becoming bullies and intimidating those students who are still interested in learning (Adelman, 1998; Ishizaka, 1998). In some extreme cases, students have committed suicide because they were unable to withstand the pressure of exams or when their results did not meet their level of expectation. Hard-working and diligent students also express their lack of sense of purpose and goal in pursuing a college degree. It seems as though their main pur-pose of study is to fulfill the goals and expectations of their family and the society.

Judging from the increasing number of student dropouts and reports of student behavioral problems, the National Council on Educational Reform, an ad hoc committee created by Prime Minister Yasuhiro Nakasone in 1983, reviewed Japan's educational system. In 1987 the council made three recom-mendations for educational reform. They called for a greater emphasis on tai-loring instructions for individual needs, shifting instructional goals from being solely academic to promote lifelong learning, and providing additional student training in computer technology (Ishizaka, 1998).

In 1996 and 1997, the Central Council on Education (CEE), an advisory committee of Monbusho, proposed several ways to meet the goals of the National Council's recommendation. These changes have an important impact on the school structure, teacher education programs, instructional pedagogy, and curriculum standards (Ishizaka, 1998).

To accommodate students with different learning styles, teachers must first learn alternative teaching methods that are most effective in stimu-lating students with different learning interests, talents, strengths, and weaknesses in their visual, audio, and linguistic skills. Students with exceptional talents and gifts will be placed into specialized programs and admitted into colleges as early as age 17. Additional full-time school guid-ance counselors will be hired to meet the growing needs of students who are less adept in a competitive learning environment (Ishizaka, 1998; Japan Ministry of Education, http://www.mext.go.jp/english/org/eshisaku/eshotou.htm).

Formal assessments will be made in the learning attitudes of students to promote lifelong learning rather than judging students solely on the basis of academic achievement. For example, integrated courses have been created and implemented on the senior high school level that allow students to design their own curriculum, set individual learning objectives, and plan their own research activities in line with their future career goals (Japan Ministry of Education, http://www.mext.go.jp/english/news/2000/10/001001.htm). By 2003 all freshmen in this course must also take another course known as Human Beings and Industrial Society. The course will give students an opportunity to reflect on their own interests, strengths, and weaknesses. Throughout the course, teachers will provide additional guidance to help them achieve their goals.

The main emphasis of the revised Courses of Study is to help students develop genuine interest in their studies, as is seen in the Japanese case studies presented by the Organization for Economic Cooperation and Development (OECD) (Japan Ministry of Education, http://www.mext.go.jp/english/org/eshisaku/eshotou.htm). Instead of teaching science and math as completely separate entities, teachers will present real-life environmental conditions for students to examine and generate their own questions (Black & Atkin, 1996). The questions they raise will serve as the primary motivation for them to engage in their own investigations. To answer their own questions, students must draw upon their own knowledge and informal experiences to develop a hypothesis and find ways to test the validity of their ideas. By giving students a window of opportunity to test their own hypotheses, they adopt a more active role in deciding and controlling their own learning experience. Most importantly, learning becomes fun and meaningful (Black & Atkin, 1996).

This type of activity fosters independent learning and encourages self-responsibility; meanwhile, it also makes the applications of science and math more concrete and realistic, when most students perceive the contrary. Second, this type of teaching methodology provides a greater equal learning opportunity for a large population of students with diverse learning styles and abilities to excel at their own pace. Third, this kind of inquiry-based activity will allow teachers to perform other forms of skill assessment on students that paper and pencil exams cannot provide.

Yet the CEE proposal was not limited to revisions in curriculum and instructional approaches. New textbooks will be devised to meet the changes in the curriculum. CEE also plans to install more computers in elementary schools to prepare students for the Information Age (Ishikaza, 1998; Japan Ministry of Education, http://www.mext.go.jp/english/org/eshisaku/eshotou.htm). Furthermore, CEE also recommended the elimination of the two half-day Saturday sessions each month in addition to 5-day classes and reduced class time to 5-day classes only. All schools are given 5 years from the date of this proposal to make the final adjustment. Thus, in two years, the number of instructional days will be slowly reduced from 240 days per year to 228 days (Adelman, 1998; Ishizaka, 1998). These reforms also indicate a need for change in the teaching and training of pre-service and inservice teachers in order to meet the newly proposed instructional objectives.

China's Education System

China is the most populous country in the world, with a population of more than 1.2 billion people. With over 200 million students attending public schools and over 9 million teachers distributed throughout the elementary, junior, and senior high schools, it is also the largest education system in the world. (Nanjundiah, 1996; Wang, 1996) Approximately 80 percent of the student population attend schools in rural areas, while the rest are enrolled in urban schools. (Changbin, 1995).

The education system is highly centralized. All teaching syllabi are designed and written by a panel of scientists and professors hired by the State (national) Education Commission. The syllabus for each subject and grade level is uniform in content and instructional objectives. Education is only partially funded by the government and local communities. Parents must pay a nominal fee for books and school uniforms.

The first 6 years of school comprise the primary grades, devoted to the development of cognitive skills, followed by an additional 6 years of high school. (Tabata and Griek, 1999) The government is currently developing a national plan for 9-year compulsory education for elementary through junior high school grades.

Students wishing to attend university must pass one of the two versions of the National University Entrance Examination. (Changbin, 1995). The quality and reputation of schools and teachers depend on the number of students passing these exams, so the requirements of this examination largely dictate the teaching and learning processes in schools as teachers tailor their lessons to preparing students for these tests. (Kwang, 2000) One version focuses on the liberal arts and the other on science. The science examination tests physics, chemistry, and biology, while the liberal arts version tests history and geography. In both versions students are also tested in politics, Chinese, mathematics, and foreign languages.

With class sizes of 40–60 students and the need to cover all of the topics required by the examinations, the main style of teaching is lecturing, with little encouragement for open discussion. Textbooks are not designed for self-learning. They consist of several volumes of approximately 150 pages in which the content is highly condensed and requires explanation by an instructor. Science material is presented in a spiral fashion throughout the middle and secondary school curricula: Information is reviewed and repeated in greater depth at each successive level. The texts also include stepwise solutions to different types of problems. Teachers assign numerous problems to prepare students for the different types of questions they may encounter in the examinations. (Changbin, 1995; Mu and Shu, 1995; Suetal, 1994; Wang, 1996; Wellington, 1992).

Many schools lack the resources to perform the necessary scientific experiments. The science curricula are very specific and highly prescriptive, and all labs must be conducted in the exact format prescribed in the curricula. Practical work in the sciences therefore consists of a few demonstrations by the teacher or by lab specialists, and labs are content-driven and are not exploratory by design. Detailed explanations of the concepts or theory behind the experiments are followed by detailed instructions on how the experiment is conducted. In the few cases where students conduct the experiments, the preparatory work is so thoroughly done that students just push a button to complete the experiment. In most cases the students' participation in experimental work is limited to recording their observations. Failure by teachers to follow the prescribed formats and content exposes their students to risk of lower passing rates in the university entrance exams, and the teacher risks reprimand or penalty of loss of extra allowances at end of term review. (Kwang, 2000)

To prepare students for the statewide exams or for districtwide competitions most teachers also supervise afterschool programs; taken with the large class sizes, teachers' workloads are greater than appears from the standard 14–16 periods per week of teaching.

Science Teacher Education Model

Japan

Universities design their own teacher-training courses, and the Ministry of Education, or Monbusho, certifies courses and provides oversight of the content of the courses and the teaching faculties of all certified universities.

About 63 percent of elementary, 43 percent of junior high, and 20 percent of high school teachers are graduates of teacher-training universities; the rest graduate from general universities. About 20 percent of elementary, 10 percent of junior high, and 3 percent of high school teachers have 2-year degrees; the rest have 4-year degrees.

Teacher education requires candidates to meet special college credit criteria. To achieve the first-class teacher certificate—held by all teachers with bachelor's degrees—elementary school teachers must have 18 college credits in their specialty subject and 41 college credits in teaching. The requirements include a minimum of two college credits each in Japanese language, social studies, arithmetic, science, life and environmental studies, music, art and handicrafts, homemaking, and physical education.

Junior and senior high school teachers must have 40 college credits in their specialty subject and 19 college credits in teaching. All prospective teachers spend 2–4 weeks in a school as part of their college training prior to certification.

The local or regional boards hire teachers. Competition for new teaching positions is intense, and the boards of education hold rigorous entrance exams and interviews for potential applicants. Teachers who pass the exam have 1 year to locate a teaching position; thereafter, they must retake the test to apply for another teaching position the following year (Adelman, 1998). The degree of competition depends on the city, level of the school, type of school, and the subject. For some subjects the ratio of applicants to positions can be as high as 30 : 1. However, in science and mathematics, the ratio of applicants to positions is much lower. The least competitive positions appear to be in technical and vocational skills, but the statistics may be deceptive since declining enrollments have been publicized and may have led to fewer applicants.

Once hired, teachers undergo a training process and various inservice training sessions during their first years in the school system. Each week, they undergo 2 days of school-based training and another day in district-level activities (Adelman, 1998). When the induction period is over, novice teachers continue to be mentored by experienced teachers in the school (Adelman, 1998). Also, teachers are normally transferred to other schools in the region on a schedule every 5–7 years.

Although elementary and junior high teachers are hired locally, prefectural governments pay half of all salaries to ensure uniformity of compensation within the prefecture. The compensation in each prefecture is based on the pay received by national school teachers, which is specified by national law.

Salary is determined by the locality, type of school, seniority, and position held in the school. In general, teachers' salaries are above the overall college graduate average, especially for female college graduates. Additionally, as civil servants, teachers are entitled to extra monetary allowances for dependents, cost of living, housing, transportation, assignments to outlying areas, administrative positions, periodic service, and diligent service. For example, in 1991, average elementary school teachers earned a base salary of $US2559 per month. But the bonuses they receive can raise this figure considerably. In 1996, the same teachers that earned this base salary received an additional $US5630 in the summer and $US6909 in the winter (Adelman, 1998).

The average length of service for elementary, junior, and senior high schools is between 15 and 16 years. The average teacher is about 40 years old; less than 20 percent are under 30, and only about 10 percent are over 55 years old.

Teachers find some differences between their expectations of what a teaching career involves and what they actually do at work. In addition to being heavily involved with committee work and attending professional development activities, teachers are expected to go on school trips and class excursions with students, take part in daily cleaning of the school, supervise clubs and other afterschool activities, and eat lunch in their homeroom with students. Elementary school teachers are expected to visit homes of students in their classrooms at least once a year, while junior and high school teachers visit homes whenever anything is amiss. Teachers are generally expected to be at their schools for at least 8 hours per day, and most teachers do all their school related work at the school, which facilitates interaction among their peers. During vacation, teachers are required to inform the school of their whereabouts so that they can be easily contacted if their students are in trouble (Adelman, 1998). Thus, Japanese teachers are not only classroom teachers, but also serve as guidance counselors to ensure the normal growth and development of a mature and responsible young adult.

In general, the system attempts to recruit, develop, and keep teachers who are motivated, and have the requisite skills and knowledge through a rigorous selection process, and upon selection, by nurturing all types of teachers through their various life stages and by facilitating sharing of professional experiences and information. Teachers are aware that the profession enjoys a high level of respect; there is intense competition to enter the profession, the salary is above average, work hours compare favorably with those in companies, and there are chances for advancement, new responsibilities, and job security.

China

In general, preprofessional training for prospective teachers follows the primary and secondary education programs for the national exams. College-level preparation programs for science teachers are tailored differently for the needs of primary, junior, and senior high school teachers.

According to the Teachers Law of the People's Republic of China and the Regulations on Qualification of Teachers, all primary and secondary school teachers must meet the following criteria: (Wang, 1996; Zheng, 2000)

- All teachers must be Chinese citizens.
- All teachers must love and commit themselves to the cause of a socialist education.
- All primary school teachers are required to attend secondary level schools to obtain their certification to teach.

Junior high school teachers are trained in 3-year colleges, while senior high school teachers are mandated to attend 4-year colleges and must attain a bachelor's degree. (Tabata and Griek, 1999; Wang, 1996) Both groups take similar courses and receive similar training. To meet the revised goals set by the New Republic, within the next few years all prospective teachers will be expected to attain a bachelor's degree.

The main goal of the science teacher education programs is to train experts in the subject area, with a solid foundation in science content and excellent problem-solving skills. (Nanjundiah, 1996) Little time is devoted to training teachers in child or adolescent psychology or in alternative teaching and learning styles and methods. Preservice teachers attend classes three times a week, from 8 A.M. to 5:30 P.M., with a 2-hour lunch break from 12:30 P.M. to 2:30 P.M. The morning sessions are devoted to content knowledge, while afternoon classes focus on conducting lab experiments. Prior to the lab, instructors explain the theory behind the experiment and provide detailed explanations of the procedures and the equipment. Students work in groups of two or three. Between 15 and 20 labs are done each semester. The courses taken are clearly delineated by area of specialization. For example, no courses in biology or chemistry are required for a physics teacher. About 50 percent of the time is spent in taking physics courses and 35 percent of the time is spent in general courses: a foreign language (usually English), political science, and mathematics. Only 15 percent of the time is spent on professional education courses such as pedagogy, general psychology, and curriculum and methods. These education courses are taken in the final semester of college. The curriculum and methods course is the most highly emphasized of these and is tailored to address the needs of high school science students. The course teaches, for the particular subject, how to prepare lab materials for experiments, set up demonstrations, and analyze the subject curriculum and textbooks, as well as time management. Methods for teaching difficult concepts are also discussed. (Nanjundiah, 1996; Wang, 1996; Xiaohui, 1991)

Student teaching lasts 6 weeks. About 20 teachers are sent to the same school, with the best students going to schools that have more qualified teachers and more resources. The methods course professor supervises these students and reviews the lesson plans before they are presented at the school. During the first 2 weeks student teachers observe an experienced teacher, study the curriculum, and return to their professors for further advice. Following this, the student teachers develop their own lesson plans, prepare their teaching materials, and submit a copy to their advisor 2–3 days before they are expected to teach. The professor makes any necessary adjustments, and copies are distributed to the regular class teacher and to fellow student teachers. (Darling-Hammond et al, 1995)

Students frequently practice the lesson by teaching to their mentor and peers and receiving their collaborative input before presenting it in the classroom. Student teachers will deliver six to eight lessons during this period, but their mentor will only closely scrutinize the first two. Thus, most of their time is spent studying the methods of the experienced teachers, rather that on actual delivery to the class. Outstanding student teachers are sometimes selected by the mentor to present their work to fellow student teachers from other groups, so an element of peer teaching and learning exists.

There are no teacher licensing exams, but teacher certification is granted after a teacher satisfactorily completes a 1-year probationary period at a local school. During the probationary period the candidate works with experienced teachers to gain experience in lesson planning and instructional strategies. Candidates are observed and reviewed by certified teachers in the subject matter area, as well as by school administrators. Upon attaining a satisfactory rating at the end of the year, the candidate becomes a certified teacher. If the rating is not satisfactory, the candidate is allowed another probationary year. If the rating is unsatisfactory after this second period, the candidate will not be allowed to teach again.

During inservice training, novice teachers work together with teachers experienced in the same discipline. This group system (*jiaoyanzu*) is built upon a hierarchical model in which experienced teachers are perceived as having great authority and deserving of a great deal of respect. The novice teachers and experienced teachers work together, study curriculum materials, and write lesson plans. The group holds weekly meetings, which offer emotional and professional support for the novices, and helps them to translate theoretical knowledge into practice by the constant interchange of useful knowledge, updated pedagogical information, and discussion of students' performance. Teachers also assist each other in testing out new instructional methods during preparation periods. The novice-mentor relationship is similar to the apprenticeship, in which beginners are expected to follow the exact regimen prescribed by the teacher and are not expected to deviate from the prescribed formats. The system helps to build strong interactive networks within the disciplines and across the larger teaching community. However, it does not support novice teachers' efforts to test unproven or creative approaches, and these teachers risk being considered as lazy and nonscholarly if their efforts are deemed ineffective.

Professional Development Issues

Japan

The program for professional development involves extensive training for novice teachers, class rotation, job rotation, peer observation and peer discussion, mentoring, outside school training experiences, opportunities to attend graduate school, research activities, and overseas travel.

All new teachers at national and public elementary, junior, and high schools and at special education schools undergo special training programs for novice teachers. All new teachers participate in 5 or more days of overnight outings devoted to training sessions in their subject area or in specific tasks such as career guidance within their school.

During their first year, elementary teachers receive 90 hours of training, 60 of which are within their schools. At this level the mix of teachers at a grade level is carefully balanced to ensure an assigned mentor to each new teacher and a group of colleagues with varying levels of experience. New teachers are required to submit detailed lesson plans, observe model classes, and participate in informal discussions. The out-of-school training includes lectures, practice teaching, volunteer work in the community, study groups, and visits to schools, educational centers for children, welfare homes, and private businesses. At the elementary level teachers often follow a group of students through two or more grades and thus gain insight into child development issues.

In their first year junior and high school teachers work a reduced teaching load of about 10 hours per week and attend the Educational Center one day each week. Training also involves visiting other schools and education-related institutions and writing extensive lesson plans. Some trainees take the role of teachers, and others take the role of students and must write lengthy critiques.

In most prefectures, teachers spend extra time in their sixth, tenth, and twentieth years at training sessions outside their school. The sessions provide time to interact with others who teach the same subjects and who are at the same career stage. Teachers in these programs are again required to write lengthy lesson plans and other reports. When a teacher becomes a grade level head teacher or advances to other administrative posts, he or she also attends training sessions. During some training experiences teachers have the chance to visit major research institutions and to see advanced laboratories and equipment. About 1,250 teachers annually are allowed to return to graduate school for 1–2 years. A program of researchers consisting of a small number of teachers with at least 17 years of experience and research students consisting of younger teachers are chosen each year by the administrators to do in-country research. The chosen teachers get time off from their school to travel to a place of their choice within Japan for a few weeks of study each year.

Nationally, about 5,000 teachers, mostly over 35 years of age, are sent abroad for training or visits to other school systems. This emphasis on visiting foreign schools as part of a lifelong education for teachers seems to reinforce a sense of professionalism and the importance of continued learning.

Another feature of the system is the job rotation of teachers. Teachers consider themselves to be hired by the board of education and not by a particular school. Teachers are transferred every 5–7 years among various schools, usually without their being consulted or having any control over their assignments. They expect this and consider it to be a normal part of teachers' careers. Teachers seem to find the variety of teaching experiences valuable.

Other Features of the Professional Development Model Some teachers voluntarily participate in teacher-run study and research groups that meet in the evenings. The meetings generally discuss new textbooks or books on teaching methods, preparation and sharing of effective lesson plans, and development of new curricula or criteria for specific subjects and grade levels. Teachers also often ask their peers to review their lesson plans or to attend a class and offer critique.

Generally, when discipline problems occur, teachers attempt to get students to resolve the issues themselves and to discuss disagreements. Teachers do not seem to feel a sense of powerlessness in the face of student misbehavior. In the few schools that have more disciplinary problems, staff changes and other structural interventions are invoked to help improve the schools.

The route to administration involves recognition beyond one's immediate school and participation in research activities. This seems to be followed more by male rather than female teachers, partly because of the time required of women for raising families.

China

Formal teacher development groups (*jiaoyanzu*) offer strong support for novice teachers. (Darling-Hammond, et al, 1995) The hierarchical model, rigidity of the curriculum, insistence on established methods, and the individual risks involved, however, are powerful factors preventing the development of creative or novel approaches to teaching. Large class sizes and the need to satisfy the curricular requirements of the national exams limit teachers to lecturing and demonstrations as the dominant teaching methods. Teachers are not supported by the system or by their mentors in efforts to conduct classes that would develop investigative and inductive reasoning skills in their students. (Darling-Hammond et al, 1995) The design of the salary ranking system does not encourage teachers to seek higher education after they attain certification. Teachers who obtain graduate degrees do not gain additional compensation, although they may find other job opportunities as a result of the additional qualifications. (Tabata and Griek, 1999; Kwang, 2000) Professional development opportunities at major research institutions are nonexistent for teachers. A few districts offer professional workshops supervised by science teachers in teacher-training institutions, and the schools pay for them, but few opportunities seem to be available (Zheng, 2000; Kwang, 2000; Tabata and Griek, 1999).

Socioeconomic/Cultural Contexts

Japan

Although Japan has a relatively homogeneous population, three factors— economic, ethnic, and regional differences—have a significant impact on the nature of schooling.

Economic Income and wealth appear to have the most striking impact on student achievement. Differences in opportunity for education deriving from economic background influence parental participation in school activities, the rate of students holding part-time jobs, and parents' and students' academic aspirations. In situations where both parents hold full-time jobs, student behavior problems and problems within families at home are more evident. Also, in wealthier neighborhoods, families could more readily afford *juku* and other extra-school academic opportunities for their children, as well as other opportunities such as overseas travel.

Regional Differences Japan strives to ensure access to high-quality public education for all students. However, in certain geographical areas, regional differences in access to school do exist. In some areas there is no well-developed system of public transportation, and children have long commutes to school. Also in such areas, the lack of large *juku* forces teachers to provide more extra classes and act as guidance counselors in helping students to pick future schools.

Ethnicity Although Japan has a relatively homogeneous population, the country is also home to several minority groups. These include the Ainu concentrated in northern Honshu and Hokkaido, descendants of Chinese and Korean immigrants, *burakumin,* and people from several different nations who have come to Japan as temporary residents to work.

The two largest groups are people of Korean descent and the *burakumin.* Over 600,000 people of Korean descent live in Japan, and there are 60 Korean schools with an enrollment of 30,000–40,000 students. The *burakumin* are ethnically and linguistically Japanese, but their ancestors were once outcasts. Despite nationwide efforts to reduce inequality, Japanese of *burakumin* descent continue to experience difficulties in access to education, jobs, and housing.

Although research on minority issues is a volatile topic, there is evidence of institutionalized discrimination faced by these students, as demonstrated by their systematically lower rates of high school graduation and college attendance when compared to the population as a whole. However, describing the barriers is difficult because many families do not wish to be identified as *burakumin,* and some schools have a strict policy of maintaining anonymity with regard to these students. (Stevenson and Lee, 1998)

CHINA

Several factors impact the science teacher in China. Confucius, the great Chinese teacher, scholar, and philosopher, is perceived as the premiere model for modern teachers to follow. His views about the teacher and about the relationships between parents, teachers, and students inform much of the philosophy of the education system. According to Confucius, if the student does not learn well at school, the responsibility is the teacher's. But if the child does not behave well, the fault is with the parents. Also, he taught that teachers are authoritative figures second in importance only to the parents. These teachings imply that the teaching and rearing of children depend on close collaboration between teachers and parents, who must form a close network of communication that also involves the child. Thus parents are charged with the responsibility to ensure that children show a high level of respect for their teachers. (Darling-Hammond et al, 1995; Wang, 1996)

Although China has vast rural, coastal, and urban areas that differ in their cultural, economic, and societal features, the centralized curriculum design feature of education system does not cater to these differences. Many of the poorer provinces are unable to supply the equipment to support the requirements of the curriculum. Recent efforts have sought to modify the curriculum to address the needs of the coastal areas, but supplementary materials to support these changes have not yet been published, and the national examination has not yet been switched to reflect the changes in these areas. Teachers therefore have little choice than to teach the old curricula. However major reform in the examination is now in progress. (Chang, 2000)

Performance of their students in the national exams is the main arbiter of teacher and school quality. However, the restrictions imposed by the system in terms of permissible teaching methods imply that the needs of students with different learning styles or rates are not addressed.

Challenges in Teaching Science

Japan

While the training system is extensive, teachers appear to have mixed views about it. Teachers perceive a need for more informal training that emphasizes human relationships and encourage the transmission of knowledge and experience from more experienced to less experienced teachers. The sixth- and tenth-year training sessions seem to provide little new information, especially of less successful approaches or significant new insights, but the exercises are useful in that they require self-reflection. Teachers seem to feel that these sessions would be more beneficial if the teachers provided more input into these sessions and if the sessions were more cooperative in nature instead of being led by elite teachers and administrators.

Teachers who resign their permanent teaching positions cannot be reinstated as permanent teachers without retaking the extensive employment examinations, regardless of experience. The extent of resignations and subsequent reentry into teaching because of life situation changes is uncertain. However, given the lower number of applicants for science and mathematics positions, the need for recertification by retaking the extensive employment exams may reduce to some extent the pool of experienced teachers in these areas. It also limits the ability of larger numbers of teachers to pursue graduate work or other professional development at their own initiative and confines them to pursuing such initiatives only within the formal system. The cultural system, which imposes the burden of homemaking on women, combined with the time demands and expectations of the profession for aspirants to administrative positions, limits the number of women who seek or attain these positions. Teachers must devote substantial additional time outside the formal teaching system to support the extensive shadow education system. To some extent this may reduce teachers' ability to pursue independent professional development.

China

Relatively little training in child/adolescent psychology and alternate teaching/learning styles implies that teachers are not prepared to address the needs of students with different learning styles. Lack of support for graduate education limits teachers' motivation to build or update their skills beyond the limited support provided within the system. The rigidity of curriculum and the need to teach to national exams discourage teachers from addressing the particular social or cultural needs of the different geographical areas of the country. Critical and inductive reasoning skills development in students is not an important teaching objective. This deters teachers from undertaking the type of exploratory exercises in which joint learning occurs and limits them to exercises in which outcomes are well established.

References

Adelman, N. (1998). *Trying to beat the clock: Uses of teacher Professional time in three countries.* ERIC Document Reproduction Service, Washington D.C. ERIC Document 420651.

Black, P., & Atkin, J.M. (1996). *Changing the subject: Innovations in science, mathematics, and technology education.* New York: OECD.

Changbin, Z. (1995). The National Examination and its influence on secondary school physics teaching in China. *Physics Education, 30,* 104–108.

Darling-Hammond, L., Sato, N., Paine, L., & Snowball, D. (1995). *Professional development: International and national perspectives. Four presentations from AAHE's National Conference on School/ College Collaboration, 1993–1994.* Washington, D.C. ERIC Document 411761.

Ishikaza, K. (1998). Reforming Japan's schools. *Principal, 77,* 24–27.

Loucks-Horsley, S., Bybee, R.W., & Wild, E.L.C. (1996). The role of community colleges in the professional development of science teachers. *Journal of College Science Teaching, 9,* 130–134.

Mu, X., & Shu, B. (1995). How teachers teach physics in China: A survey. *Physics Teacher, 33,* 45–47.

Nanjundiah, S. (1996). The training of science teachers in the People's Republic of China. *Journal of Science Education and Technology, 5,* 161–65.

National Institute on Student Achievement, Curriculum and Assessment. Office of Educational Research and Improvement, US Department of Education. Washington, D.C. *The educational system in Japan: Case Study Findings.*

Petterson, L. (1993). Japan's "cram schools." *Educational Leadership, 50,* 56–58.

Su, Z., et al. (1994). Teaching and learning science in American and Chinese high schools: A comparative study. *Comparative Education, 30,* 255–270.

Tabata, Y., & Griek, L. (1999). *Ensuring opportunities for the professional development of teachers. Innovation and reform in teacher education for the 21st century in the Asia-Pacific region. Seminar Report.* UNESCO, Bangkok, Thailand. ERIC Document 430913.

Wang, W., (1996) Science education in the People's Republic of China. *Science Education, 80,* 203–222.

Wellington, J. (1992). Physics teaching and teacher training in China: A Western perspective. *Physics Education, 27,* 130–133.

Xiaohui, Y. (1991). Teaching biology in China. *Journal of Biological Education, 25,* 141–145.

Internet References

Japan Ministry of Education
http://www.kantei.go.jp/foreign/link/link3.html#21

http://www.mext.go.jp/english/news/2000/10/001001.htm
http://www.mext.go.jp/english/statist/index.htm

Personal Interviews

Kwang, T.C. Personal interview. 26 July 2000.
Zheng, C.X. Personal interview. 26 July 2000.

Chapter 4
Professional Development for Teachers of Science
The Canadian Experience

PIERCE FARRAGHER

History of Science Education in Canada

Canada is a confederation consisting of 10 provinces and three territories. Its population is currently 30 million. A parliamentary democracy ensures its citizens are represented in an elected House of Commons. The government is headed by the Prime Minister, and areas of government jurisdiction are divided between the federal government and the provinces (Taylor, 1997). The provinces and territories are responsible for education and while educational structures and institutions across the country are similar in many ways, they have been developed by each jurisdiction to respond to the particular circumstances and historical and cultural heritage of the population they serve. They also reflect the circumstances of regions separated by great distances.

Canadians place a high priority on education and training, and spending on education is the second largest public expenditure (approximately 20% of all public expenditures) after health care.

Canada, being the second largest country in the world, offers great diversity in political organization, geography, and cultural makeup. These factors have greatly influenced the structure and nature of its education system. A study by Connelly, Crocker, and Kass (1985) stated that, "to understand science education it is important to understand the various historical, political, geographic and philosophical factors that have shaped Canadian education and thus provide the milieu for Canadian science education."

Canada is officially bilingual (English is the mother tongue of about 59% of Canadians, French is the first language of 23% of the population). In 1982 Canada adopted a new constitution, which guaranteed the right to education in both languages. Language problems arise in science education since few science teachers can teach in minority languages. The range of francophone science textbooks and teaching resources suitable for use in the Canadian context is also limited. This poses problems for curriculum developers in many jurisdictions.

Over the past 50 years or so, the federal government has transferred revenues to the provinces to support postsecondary and vocational programs, to promote bilingualism and Canadian studies, to provide financial support to needy students, to encourage research, and to help remove economic disparities among provinces.

In the 1960s and 1970s science facilities and equipment were improved through federal funds for capital expenditures. The federal Metric Commission provided assistance for new metric textbooks. New curricula and resources were introduced, especially in relation to the environment and the use of energy and resources. Assistance was provided in many cases by federal and provincial ministries other than education ministries.

In the 1980s and 1990s, economic constraints and declining enrollments resulted in funding cutbacks, teacher layoffs, fewer new hirings, closure of small schools, larger classes, and the phasing out of low-enrollment classes. Science has not escaped these problems, and science teachers are often called on to teach subjects for which they are not qualified. A reversal of this trend seems to be occurring at the start of the new millennium, and the prediction is for a shortage of teachers in all subject areas (particularly in the sciences) over the next 5 years.

Canada has a highly sophisticated communication system, and education is one of the beneficiaries of this system. Correspondence courses have long been available in science subjects. Radio and TV have played a prominent role in the production of some science-related courses and programs. Today, science educators employ emerging technologies in their work such as computers, interactive video, CD-Rom, satellites, audiovideo teleconferencing, e-mail, and the Internet.

Science education in Canada has traditionally focused on practical issues. During the 1960s, the Economic Council of Canada advocated that education be viewed in terms of its contribution to human productivity. Science was promoted as the means to securing a better economic future for Canada. Science courses were developed for academically oriented students who were bound for universities. These courses, however, were often out of reach of the general student. Since the 1980s new academic courses have been developed for university-bound students, inquiry-based courses for students proceeding to community colleges or to employment, and basic courses to equip students with skills to cope with everyday life.

While the science curriculum today continues to change, the main emphases are on practical applications, wise use of resources, ethical uses of scientific knowledge, and skills for living in a high-tech society.

Funds have been provided by the federal government to develop Canadian content textbooks and other resources. In science, "Canadian" content refers to matters such as environmental concerns, the use and conservation of local resources, the effects of science and technology on Canadian society, and the study of Canadian scientists and scientific accomplishments. To date, however, most of the science textbooks used in Canadian schools are produced in the United States and have rather limited Canadian content.

Structure of Education and Training in Canada

Pre-elementary Programs

Most provinces and territories offer preschool programs or kindergartens that are operated by the local education authorities, providing 1 year of pre-grade 1 education to 5-year-olds.

Elementary and Secondary Education

Public education is provided free to all Canadian citizens and permanent residents until the end of secondary school, normally at age 18.

Elementary schools in most jurisdictions cover the first 6–8 years of compulsory schooling. Afterwards, children proceed to a secondary education program. A great variety of programs—vocational (job training) as well as academic—are offered at the secondary level. Secondary school diplomas are granted to students who pass the compulsory and optional courses of their programs.

The point of transition from elementary to secondary school may vary from jurisdiction to jurisdiction. The elementary-secondary continuum can be broken up into schools that group, for example, kindergarten to grade 6, grades 7 to 9 (junior secondary or intermediate), and grades 10 to 12 (senior secondary). In Quebec, secondary schooling ends after 11 years of study.

Postsecondary Education

Once secondary school has been successfully completed, students may apply to a college or a university, depending on the jurisdiction and on whether they qualify. Postsecondary education is available in both government-supported and private institutions, some of which award degrees. Today in Canada, some 200 technical institutes and community colleges complement about 100 universities, attracting approximately 1 million students.

Number of Educators at the Elementary and Secondary Levels

The number of full-time educators in public schools in Canada has increased consistently since 1984–1985. It is interesting to note that, while the number of male elementary and secondary educators has remained fairly constant over time, the number of female educators has increased consistently.

Women continue to make up the majority of educators employed in elementary and secondary schools.

Gender equity may slowly be gaining ground in the teaching profession. Much ground is yet to be gained, however, considering that women constitute 61 percent of all full-time educators. There is a need for proactive recruiting to provide for a more even distribution of teachers in particular secondary areas for example, in mathematics and science.

Preservice Science Teacher Education

Certification of Teachers

Canadian teachers follow one of two certification programs: a 4-year university course leading to a bachelor of arts or science degree, plus 1 year of teacher education, or a bachelor's degree followed by 2 years of teacher education. Faculties of education within universities conduct teacher education, and the Ministry or Department of Education grant teaching certificates.

To qualify as an elementary school teacher, a candidate must complete between 4 and 6 years of postsecondary training. Most of the time spent on training in pedagogy focuses on general applications across curricular areas. While some universities require at least one course in mathematics methodology, others do not.

At the secondary level, requirements for teacher certification range between 5 and 6 years of postsecondary training. The number of required pedagogy courses specific to science or mathematics vary from none to one or two. Teachers of secondary science or mathematics, however, are expected to have an academic major in the area, which includes completion of at least three or four subject matter courses. To increase their job prospects most student teachers are encouraged to present two teaching areas, for example, biology and physical education, physics and mathematics, biology and chemistry, biology and english, and so on.

Inservice Science Teacher Education

Inservice development and upgrading for science teachers is usually done through courses offered by departments of education, school boards, or faculties of education. Most school boards allow up to five PD (professional development) days per school year for inservice development. Many teachers now avail of the large number of courses offered through the Internet. Courses and workshops are also offered by teachers' associations and other professional organizations. With respect to graduate work, teachers may pursue an M.Ed. degree in science education or curriculum studies after 2 years of teaching experience provided that GPA requirements are met. An M.A. degree may be pursued at any time by qualified applicants.

Due to curriculum changes and reassignment of teaching subjects, additional training is essential for most science teachers. Areas of focus include computer applications and new technologies, alternative teaching strategies such as cooperative learning, the expansion of student evaluation methods such as observational techniques, criteria-referenced evaluation, student-parent-teacher conferences, and portfolio assessment. The additional training can ultimately lead to higher certification levels with corresponding salary increases and potential promotion to department head or other administrative positions within the school.

Teacher Profile

The average age of Canadian teachers is increasing. Predictions are that a large number of teachers will be retiring or nearing retirement in 2001–2010. All indications are that large numbers of teachers, both elementary and secondary, will be required over the next 10 years. The prospects for secondary science teachers looks especially bright. Physics teachers, especially female, continue to be in demand, while the demand for chemistry teachers is also good.

While women constitute the majority of full-time teachers in Canada, at 61 percent, they are less likely to teach science and mathematics than men. Robitaille (1997) suggested that some incentive might be provided for women to teach science and mathematics.

Research on Inservice Science Teacher Education in Canada

Canadian teachers are being urged to pursue career-long professional development that moves beyond skills training and generic inservice models to a more flexible engagement with other experts in the field (Ontario College of Teachers, 1998). Research on teacher professional development (TPD) suggests that teachers need more opportunities (1) to access and discuss exemplary reform-based materials, (2) to co-construct and publish resources for new teaching practices, and (3) to collaborate on the creation of locally relevant solutions by participating in a professional community of practice (Lieberman & McLaughlin, 1995; Loucks-Horsley, Hewson, Love, & Stiles, 1998). Such opportunities, which require a commitment to sustained professional education, have become virtually nonexistent in Canada for a number of reasons, including changes to funding structures, administrative reorganization, and a large number of retirements. A review of the recent literature suggests that the predominant forms of professional development (the workshop model and the train-the-trainer model) are not sustained, generative, or collaborative and do little to generate or sustain significant change to teacher practice and professional beliefs (Loucks-Horsley et al., 1998).

While the number of research reports on inservice science teacher education in Canada is limited, the work of Connelly, Crocker, and Kass (1985) and Wideen and Holborn (1990) is most relevant. The latter state that most practicing teachers in Canada have access to a broad range of opportunities for inservice education. Fox (1985) identified the various formats that PD take in Canada; this included workshops; conferences; media sources; in-school PD; board-level PD; regional PD; summer programs (institutes); full-year activities (sabbatical); academic courses in the subject area, related areas, or in education offered by universities and faculties of education; and workshops sponsored by publishers and equipment manufacturers. Individually, teachers usually choose their own professional development activities, even though goals and guidelines for change may be set by a larger organization such as the provincial Ministry of Education, a school district, or a school staff. An increasing emphasis has been placed on long-term goal setting and collaborative efforts toward educational change through school staff development projects, districtwide inservice projects, and provincially guided curriculum development and implementation programs (Wideen & Holborn, 1990).

Collaborative inservice projects involving school districts, universities, and Ministries of Education have been developed in several Canadian provinces. The reason studies on inservice science teacher education in Canada are scarce and difficult to locate, presumably, is because the majority are never published. Writing for publication is not a high priority for school districts who carry out investigative projects. The qualitative nature of inservice research differentiates it substantially from research on preservice and induction. The challenge of helping teachers accept and embrace educational change remains an ongoing concern. Most agree that effective inservice must go beyond short-term "quick-fix" approaches and that teachers must have some ownership over the change process (Wideen & Holborn, 1990).

Hills (1990) reported on a 2-year program of inservice education in science for elementary school teachers. The project focused on enabling teachers to arrive at a fuller understanding of the nature of science and of scientific inquiry itself. A selection from the end-of-year interview transcripts helped to evaluate the project from a teacher's perspective.

McFadden (1991) conducted a survey of grade 7, 8, and 9 science teachers in New Brunswick and Nova Scotia as part of a research program to determine the consequences for teaching and learning of the recent introduction of the SciencePlus program developed by the Atlantic Science Curriculum Project and the need for further curriculum and professional development. The survey explored some of the conditions of teaching, teachers' goals and instructional practices, and teachers' professional development needs and preferences.

A study by Vey (1992) explored secondary science teachers' perceptions of and concerns about the development and implementation of a proposed Science Technology and Society (STS) course in the schools of Newfoundland and Labrador. The study provided insight into how science teachers perceive STS topics in the curriculum and gave teachers an opportunity to have input into the development of the STS course and the nature of the inservice required.

Research by Hopkinson (1993) entailed an exploratory qualitative analysis of the effectiveness of an inservice science program for elementary teachers, specifically focusing on the experiences of some of the participants. The program consisted of a 1-week institute, conducted at Science World, Vancouver, British Columbia, and a subsequent field-based implementation course offered by Simon Fraser University. Results indicated that teachers perceived a dramatic shift in their science teaching because of their experiences in the program. Teachers expressed an appreciation of a constructivist perspective on the learning and teaching of science in the classroom.

McTavish (1993) collected data in personal interviews with Ontario physics teachers. He identified six sources of influence on changes in evaluation practices: personal beliefs, other teachers, students, the school, the Ontario Ministry of Education, and professional development activities. The strongest influence for change in evaluation practices is clearly the Ontario Ministry of Education. Other teachers, the school, the students, and professional development together made up just over 25 percent of the references to sources.

Brooks (1995), in response to concerns within the science education community about the status of science in elementary schools, conducted a survey at the Elementary Science Leadership Forum held at the University of British Columbia. The survey explored forum participants' perceptions of needs, resources, and teacher inservice in their districts as related to the teaching and learning of science.

The respondents' most pressing needs to improve the teaching and learning of science in elementary schools were teacher inservice opportunities followed by funds for science equipment and expendables. Many respondents felt that science was being subsumed by other disciplines, especially language arts. They also felt that equipment and supplies for teaching science are inadequate or poorly organized, that many elementary schools lack the physical space for teaching science, and that poor communication exists between the district office and the classroom teacher.

When asked, "Which of the following topics for science inservice are commonly requested by elementary teachers or administrators in their district?" the leading topics were:

- Using instructional strategies in science
- Using supplementary programs
- Integrating science with other curricular areas

Martin, Miller, and Szostak (1995) reported on the proceedings of a Geoscience Workshop held to familiarize New Brunswick teachers with the subjects of geology, mining, and mineral development. The workshop approach focused on tools and resources that teachers could use in the classroom.

An article in the *Globe and Mail* newspaper dated August 11, 2000 (Procuta, 2000), described a summertime science workshop conducted for elementary teachers in the Toronto area. Almost 6,000 Ontario elementary and secondary teachers took weeklong workshops in the summer of 2000 to familiarize themselves with the new curriculum introduced into Ontario schools in 1998. Demand was so high that extra sessions had to be added. The workshops were organized by teachers' unions and paid for by the Ministry of Education.

While teachers were enthusiastic about the new hands-on science curriculum, many worried about the problem of having to teach science on a shoestring. They expressed concern about scrounging for even the most basic materials to meet the demands of the new, more complex science curriculum—items such as microscopes, thermometers, mirrors, and goggles. Teachers indicated that boards of education did not have enough money to buy the resources needed.

It is clear from the research that Canadian Science teachers, both elementary and secondary, value inservice and see it as essential to further their professional development. It must be recognized as a part of the job and be a continuous, ongoing process done every week over long terms to be really effective. It must involve all science teachers.

Science Curriculum and Pedagogy

Goals for the Science Curriculum

Common goals for science are depicted in the publication *Science for Every Student: Educating Canadians for Tomorrow's World*, published by the Science Council of Canada (1984). In this report, the Council stated four broad aims for all students.

- To encourage full participation in a technological society
- To enable further study in science and technology
- To facilitate entry into the world of work
- To promote intellectual and moral development of individuals

These broad aims or goals are reflected in recent science curricula through increased emphasis on understanding scientific processes and principles, conceptual understandings approached from a phenomenological basis, integration within the sciences, a focus on the impact of technology on society with respect to environmental and resource issues, and the promotion of science-related careers.

Major Changes in the Science Curriculum

There has been a continued emphasis on both science processes and on knowledge and skills during the past 10 years. The focus on processes such as observing, classifying, inferring, and communicating has begun at lower grade levels, moving toward a greater emphasis on knowledge and skills in the upper grades.

A shift has occurred, however, in the content and method of delivery of the science curriculum, grounded in changes in philosophy and the changing needs of society. They have included the following:

- Understanding scientific processes and principles
- Solving problems rather than memorizing facts and performing traditional laboratories
- Conceptual understanding approached from a phenomenological rather than a mathematical basis
- Integration within the sciences
- Integration of sciences within other subject areas
- Issues surrounding the impact of science and technology on society, particularly with respect to environmental and resource issues
- Teaching from constructivist principles
- Issues of gender equity and cultural bias

There has been general acceptance of the use of calculators and computers by students and teachers of science at all levels of the system. The availability and use of audio, video, and software materials have also increased significantly. Technology has also had an impact on the nature of courses offered in science. For example, new science and technology courses have been developed in several provinces for use at the secondary level.

The Third International Mathematics and Science Study

Since 1959 the International Association for the Evaluation of Education Achievement (IEA) has conducted a series of international comparative studies. The Third International Mathematics and Science Study (TIMSS) of 1990 was the largest of these studies.

Key Findings

Grades 3–6 Science
- At the grade 3 level Canada was above the international average, in ninth place out of 24 countries.
- At the grade 4 level Canada was again above the international average, in ninth place out of 25 countries.
- Gender differences at the third and fourth grade levels were much less pervasive than at the seventh and eighth grade levels.
- In Canada, the majority of grade 4 students were taught science by female teachers.

Grade 7–8 Science While Canada performed above the international average in science achievement at both the third and fourth grade levels, it was just at the average at the eighth grade level. Singapore was the top performing country in science achievement at both the seventh and eight grade level. Interestingly, Canada outperformed Singapore at both the grade 3 and grade 4 levels.

A report released in December 2000 showed Canadian grade 8 students near the top of the class in international math and science tests from 38 countries. In science, Canada had a mean result of 53 compared to a high of 57 for Taiwan and Singapore. The lowest mean was 24. Canadian students made much greater use of calculators in the classroom than students from most other countries did. Gender differences were quite pervasive, with boys outperforming girls, particularly in physics, chemistry, and earth science.

Grade 12 Mathematics and Science Literacy Test (21 countries)
- Canada performed above the international average.
- Canada had a higher achievement in science literacy than in mathematics literacy.
- Canadian males had significantly higher average achievement than females in science literacy.
- Canadian students reported positive perceptions about their performance in science, with more than 70 percent agreeing that they usually did well.
- In every country, final-year students whose parents had more education had higher science literacy. More than 30 percent of Canadian students indicated that at least one parent had finished university.
- The frequent use of calculators was positively related to science literacy in all countries. Students who reported the most calculator use on the test performed best.
- More than 50 percent of Canadian students reported at least weekly use of computers.
- At least 25 percent of Canadian students reported working for 3 hours or more each day at a paid job.

Grade 12 Physics Test (16 countries)
- Canadian males had significantly higher physics achievement than did females. The proportion of males and females in Canada taking physics were about equal.
- Canadian students performed relatively less well in mechanics and relatively better in heat than they did on the physics test as a whole.
- More than 25 percent of the Canadian students reported conducting laboratory experiments in most or all physics lessons, in contrast to Germany, Austria, and Greece, where the majority of students reported that they never or almost never conducted laboratory experiments.

There was no consistent relationship between frequency of conducting laboratory experiments in class and physics achievement.

- The plans for Canadian students who had taken final-year physics centered mainly on university. The most popular areas of study were engineering, mathematics or computer/information sciences, health sciences or related occupations, and business. Males often outnumbered females by a substantial margin in engineering and in mathematics or computer/information sciences.

The Pan-Canadian Protocol for Collaboration on School Curriculum

This protocol was adopted in February 1995 by the Council of Ministers of Education of Canada. It acknowledged that education is a provincial responsibility and that interjurisdictional cooperation could contribute to improving the quality of education in each jurisdiction.

The intent of the protocol was to improve the quality of science education in Canada. The Common Framework of Science Outcomes was the first joint development project undertaken as part of the protocol. It identified beliefs about science education, outlined general and specific learning outcomes, and provided illustrative examples. It provided common ground for the development of curriculum expectations within each participating jurisdiction. The framework provided more consistency in the learning outcomes for science across jurisdictions, enabling increased transferability for students moving across Canada.

Science for Canadian Students

Canadian society is currently experiencing rapid and fundamental economic, social, and cultural changes that are affecting the way we live. Canadians are also becoming aware of an increasing global interdependence and the need to sustain a shared environment and its resources. The emergence of a highly competitive and integrated international economy, rapid technological innovation, and a growing knowledge base will continue to have a profound impact on our lives. As advancements in science and technology play an increasingly significant role in everyday life, science education will be a key element in building a strong future for Canada's young people.

Future Directions of Science Education in Canada

The future of science education in Canada is partially outlined by the Council of Ministers of Education's (1997) Pan-Canadian Framework for Science and the continuing shortage of science teachers (L.D. Yore, personal communication, May 31, 2000). The framework attempts to outline the common vision, foundations, and learning outcomes for science across the nine provinces (excluding Quebec) and the three territories to facilitate the movement of students across these jurisdictions and to improve the quality and costs of instructional resources. Like many other countries, Canada has described a vision of science literacy for all citizens that involves an understanding of the interrelationships among science, technology, society, and the environment; the big ideas from science; science skills and processes; and attitudes toward science. The framework implies a constructivist orientation to science teaching and encourages the use of alternative assessment strategies. Assuming that the framework maps the emphases for the next few years, it becomes apparent that the science education community needs to find ways to increase the number of science teachers being educated and revise the teacher education and professional development programs to emphasize constructivist teaching strategies and alternative assessment techniques. Increasing the number of preservice teachers is made more difficult with the current demand for science graduates by a technology-oriented economy that recruits most of the B.Sc. graduates produced by Canadian universities, leaving very few for post-B.Sc. teacher education programs and the teaching profession. The revision of teacher education programs and professional development of practicing teachers is made more challenging in Canada by strong teacher unions and certification policies that promote life certificates without subject specialization. As in many other countries, educational research and theory have had little influence on classroom practice and professional beliefs.

References

Brooks, C. (1995). Elementary science leaders' perceptions of district needs resources and teacher inservice in British Columbia. *Catalyst, 38*(2), 19–28.

Connelly, F.M., Crocker, R.K., & Kass, H. (1985). *Science education in Canada, volume 1*. Toronto: The Ontario Institute for Studies in Education.

Council of Ministers of Education, Canada. (1997). *Pan-Canadian protocol for collaboration on school curriculum: Common framework of science learning outcomes K-12 (draft)*. Toronto.

Fox, D. (1985). Professional development: Draft position paper of the Science Teachers' Association of Ontario. *Crucible, 16*(3), 14–17.

Harmon, M., Smith, T.A., Marlin, M.O., Kelly, D.L., Beaton, A.E., Mullis, I.V.S., Gonzalez, E.T., & Orpwood, G. (1997). *Performance assessment in IEA's Third International Mathematics and Science Study (TIMSS)*. Chestnut Hill, MA: Boston College.

Hills, G. (1990). *Innovative inservice in elementary science: Final report to the Ontario Ministry of Education on a two-year program of inservice education in science for elementary school teachers, 1998–89.* Kingston, ON: Frontenac County Board of Education, Frontenac-Lennox and Addington County Roman Catholic Separate School Board; Toronto: Ministry of Education.

Hopkinson, P. (1993). *Science in the most curious places: An analysis of an in-service science education program for elementary school teachers.* Unpublished dissertation, Simon Fraser University, Burnaby, British Columbia, Canada.

Lieberman, A., & McLaughlin, M. (1995). Networks for educational change: Powerful and problematic. In M. McLaughlin, & I. Oberman (Eds), *Professional development in the reform era* (pp. 673–677). New York: Teachers College Press.

Loucks-Horsley, S., Hewson, P.W., Love, N., & Stiles, K.E. (1998). *Designing professional development for teachers of science and mathematics education.* Thousand Oaks, CA: Corwin Press.

Martin, G.L., Miller, R., & Szostak, J. (1995). *Helping teachers teach: Report on the Geoscience Workshop.* Fredericton: Natural Resources and Energy, Minerals and Energy.

McFadden, Charles, P. (1991). *Association of selected teaching conditions with reported instructional practices: From a survey of New Brunswick and Nova Scotia grades 7, 8 and 9 science teachers.* Atlantic Science Curriculum Project. Fredericton: University of New Brunswick Atlantic Science Curriculum Project.

McTavish, James F. (1993). *Factors influential in changing the practices of Ontario physics teachers for evaluating student achievement.* Unpublished dissertation, University of Toronto, Toronto, Ontario, Canada.

Ontario College of Teachers. (1998). *Standards of practice for the teaching profession.* Toronto: Author.

Procuta, E. (2000). Summertime workshops help teachers teach. *Globe and Mail,* August 11, p. A15.

Robitaille, D.F. (Ed.). (1997). *National contexts for mathematics and science education: An encyclopedia of the education systems participating in TIMSS.* Vancouver: Pacific Educational Press.

Science Council of Canada. (1984). *Science for every student: Educating Canadians for tomorrow's world.* Report 36. Ottawa City. Publisher: Government Publications of Canada. Author.

Taylor, A.R. (1997). Canada. In D.F. Robitaille (Ed.), *National contexts for mathematics and science education: An encyclopedia of the education systems participating in TIMMS* (pp. 70–80). Vancouver: Pacific Educational Press.

Vey, Bruce W. (1992). *Proposed science, technology and society course for secondary schools in Newfoundland and Labrador: Teachers' perceptions and concerns*. M.Ed. thesis, Memorial University of Newfoundland.

Wideen, M.F., & Holborn, P. (1990). Teacher education in Canada: A research review. In R.P. Tisher & M.F. Wideen (Eds.), *Research in teacher education: International perspectives* (pp. 11–32). London: Falmer Press.

Chapter 5
Changing Traditions of Science Teacher Professional Development in the Philippines

PURITA BILBAO, LOURDES N. MORANO, TESSIE
BARCENAL, MERILIN A. CASTELLANO, SHARON NICHOLS,
AND DEBORAH J. TIPPINS

This chapter is the work of a collaborative team of science teacher education researchers from the United States and the Philippines. Our work together was initiated in response to results of the Third International Mathematics and Science Study (TIMSS) (Beaton, Martin, et al., 1996; Beaton, Mullis, et al., 1996). According to the TIMSS, Filipino students performed poorly in both mathematics and science. Criticisms have been raised regarding the ways these results have been interpreted. Bracey (2000) asserts that the systems and cultures of the nations involved in the study differed to the extent that "renders the scores uninterpretable" (p. 4). What do international comparisons of science teaching and learning tell us about the contexts in which studies such as the TIMSS take place? How do cultural practices and language diversity shape science teaching and learning in classrooms? What sorts of historical legacies and political struggles have helped to shape what we see in today's classrooms? How is science education viewed in light of other forms of traditional knowledge in communities? Accordingly, these were the sorts of questions we wanted to explore as we worked to envision professional development that would enhance science education relevant to the experiences of teachers and students living in the Philippines.

In this chapter, we look at historical changes that have taken place in Filipino science education and teachers' professional development. The legacies of these historical events are seen through the present-day narratives of teachers about their classroom science experiences. We use a narrative perspective in our work because it reflects a contextualized view of teachers' thinking and their practice. Clandinin and Connelly (1996) have described the contextual nature of teachers' professional knowledge as "landscapes" that have "exceedingly complex places with multiple layers of meaning that depend on individuals' stories and how individuals are positioned on that landscape, as well as the landscape's own narrative history of shifting values, beliefs, and stories" (p. 30). To help illustrate the landscape of science teacher professional development in the Philippines, we begin with two cases written by Filipino elementary science teachers to contextualize the historical overview of science teacher learning presented in later sections of this chapter. We also describe future directions being explored for professional development in the Philippines, including our own approach using case-based pedagogy. Concluding comments discuss how historical and contextual aspects of professional development hold important implications for global science education reform.

Contextualizing Professional Development in the Philippines: Cases of Elementary Science Teaching

Over the past 2 years, our research partnership has centered its focus in the port city of Iloilo on the island of Panay in the Philippines. The city is located in a rural region where a majority of citizens practice rice, mango, and coconut farming. Panay is one of 7,100 islands that comprise the Republic of the Philippines; on any given day the number of islands change depending on the ebb and flow of tides. In our most recent inquiry we have explored science teaching and learning through the development of a case-based curriculum for science teacher education. Participants in this study included six prospective elementary teachers, their respective "critic" (cooperating) teachers, and a research team comprised of four Filipino science educators from West Visayas State University (WVSU) located in Iloilo City. Two science teacher educators from universities located in the Southeast region of the United States joined the Filipino researchers at WVSU to participate in collaborative research. The two U.S. science educators, like their Filipino colleagues, were interested in collaborating to explore critical issues of elementary science teacher learning and case-based pedagogy and alternative research approaches.

Inquiring through Case-Based Pedagogy and Narrative Research: Our Theoretical Orientations

The guiding theoretical perspectives for our work are drawn from case-based pedagogy and narrative research approaches being used in teacher education. Case-based pedagogy has become increasingly appreciated as a professional development approach for science teachers (e.g., Abell et al., 1996; Howe & Nichols, 2001; Kagan, 1993; Koballa & Tippins, 2000; Loucks-Horsley, Hewson, Love, & Stiles, 1998.) A case, in most simple terms, can be thought of as a "story" that features a dilemma written by a teacher to represent the multiple and complex ways they interpret their science teaching experiences. In this study, we have restoried our own experience by rereading and talking across the multiple data sources and our diverse interpretations. Case-based pedagogy was an expression we used early in our work as a reflective strategy involving teachers writing and discussing their cases with each other. Only after we engaged study participants in case writing and sharing did we see case sharing as a powerful *experience*. Case-based pedagogy shifted from being a tool for reflection and became a pedagogy of experience (see Arellano et al., 2001a,b). The research team had also considered their assumptions toward conducting research on using case-based pedagogy. We had identified action research as a methodological orientation that would embrace a collaborative spirit of inquiry as a mutual process for generating learning among study participants and the research team. While the research team approached the project with a specific strategy for professional development in mind (i.e., using case-based pedagogy), we looked to the teachers to bring the knowledge and questions that would drive our inquiry and learning. Accordingly, we identified narrative inquiry as a form of action research that contrasted *formalistic research*—a tradition of inquiry that has tended to give priority to using theoretical frameworks and the production of results as exemplars of formal categories (Connelly & Clandinin, 2000). Narrative inquiry, as way of understanding experience (Bruner, 1986; Polkinghorne, 1988), allowed the research team to live, tell, relive, and retell stories as a means for exploring our notions of case-based pedagogy. Accordingly, the narrative research orientation allowed us to create research texts that represent the interpretations and experiences of our study participants as well as our own (e.g., Arellano et al., 2001b).

A Virtual Visit to Elementary Classrooms in Iloilo through Two Cases

Two elementary teachers who teach science at the Integrated Laboratory School located at WVSU wrote the following cases. We selected these two cases because they provide glimpses of experiences elementary Filipino teachers described about preparing for and teaching science in their classrooms. We begin with a case written by Ms. Rossini Monsalud, a teacher of science for grades 1–3.

Rossini's case represents several issues that concern the professional development of elementary science teachers. Her feeling of apprehension toward science is a common experience to elementary teachers worldwide. Likewise, her reliance on texts to provide her with science knowledge for teaching science is also a strategy often seen among elementary teachers. Like many elementary teachers, Rossini perceived herself as lacking knowledge to teach science and believed that accurate facts should be obtained from books to compensate for her deficiencies. Rossini viewed science as a factual body of knowledge and science pedagogy as the transmission of facts

---

Case 1: Teaching about Planets: A Universe of Confusion in Texts, by Ms. Rossini G. Monsalud

I am Rossini, Adviser of Grade II-Masayahin. Science is not my field of specialization, however, I love teaching science since this was my favorite subject when I was still an elementary grader. When I was a sophomore college student, I specialized in Music, Art, and Physical Education (MAPE). (I can say that music is a science of sound, though part of it is of aesthetic concern.) After graduation, I taught in one of the private schools in Iloilo City teaching MAPE subjects. However, when I transferred to WVSU-Integrated Laboratory School, the principal let me teach science in three grade levels and that's the time I felt butterflies inside my stomach. My problem was that I had a little idea of the concept, strategy, and processes to teaching science. So I started reading science encyclopedias, and other science books aside from the textbook. I was very excited, anxious, afraid—it's just a mixture of feeling. Prominent in my plan was to ask help from my fellow teachers who are science specialists.

The first week was a little rough. Children are to be reminded of their behavior during group work, how to clean assigned places, and how to make sure of their thinking skills during the activity. As for me, I have to answer some questions accurately regarding the topic, which is confusing, and to spend my time and effort reading.

One time, our lesson was about "The Nine Planets." We were discussing about the biggest, the nearest planet to the sun, the farthest, and the smallest planet. A child raised his hand and told us that the smallest planet is Mercury though our textbook says it's Pluto. I was quiet for a while and let them talk about it for some time. The last resource I did was to let them bring encyclopedias the following day regarding the topic. I went to the library after class and got all the books regarding planets. I was again confused since some books say Mercury is the smallest and others say Pluto is smallest.

---

to young learners. Accordingly, some problems of elementary science teaching are not unique to the Philippines.

As we pondered why some problems of elementary science teaching might be so prevalent in elementary classrooms around the world, we began to consider what sorts of professional development practices might be commonly contributing to this problem. Another case, written by Tomasa Ferando, a third-grade teacher, provided insights regarding this question.

Case 2: Room on Fire! by Prof. Tomasa M. Ferando

I'm a grade 3 teacher. I've been in this grade for more than 10 years already. Though my line of specialization is reading, I cannot go away from teaching science because it is said that if you're an elementary grade teacher, you are expected to teach all subject areas, and science is one of those. Despite all the apprehensions I tried my best to cope with the expectations. I took pains in really studying, and reading science books in order to get concepts for my 40–50 pupils to learn everyday.

As I continue my role as a science teacher, I learned to love teaching science. At first I merely used the "Chalk talk" way of teaching, but the administration was so supportive that they sent me to attend a national seminar-workshop at the University of the Philippines Diliman, Quezon City, Philippines. There, I learned a lot of strategies— one of which was a Hands-On Minds-On strategy, a practical work approach in teaching science. When I came back, I applied the strategy to my pupils. Since my grade 3 pupils were neophytes with regard to the use of science equipment, I started introducing some to them. I grouped them, and I made 10 groups, five pupils in each group.

Days go on, and all hands-on activities were done smoothly with active participation among my pupils. My dilemma came when one day, our lesson was about testing of sugar in starch food. Of course, we need to use an alcohol burner to boil the solution. I gave instructions and precautions on how to use the alcohol lamp so that the alcohol will not spill and scatter on the floor. (We don't have a laboratory where to perform science experiments. So, we just make use of all the vacant spaces inside the classroom.) One of the groups was very curious that they forgot all about my precautions and they tilted the alcohol lamp after several attempts of lighting it with the matchstick. After a while, as I as supervising the other groups, one of my pupils shouted, "Ma'am, Fire! Fire!" All the pupils were shouting, yelling. The room is in chaos. My adrenaline rose up. My whole body trembled. No word came out of my mouth. I didn't know what to do. I grabbed a pile of United Nations flags and place these over the fire. That was the time when I was relieved of my tension/nervousness. I reprimanded the group for not following my instructions; however, instead of showing fear they still enjoyed looking at the fire spreading on the floor.

Tomasa Ferando, like her colleague Rossini, was uncomfortable with the idea of teaching science. She too resorted to books to provide her with the knowledge and activities she needed to teach science. Her case also indicated that teachers in Iloilo sometimes must attend workshops located far away from their own province, travel that frequently involves overnight ferry transportation. When she attended the workshop, she learned an alternative pedagogical strategy that featured hands-on science (practical work) activities versus the more traditional practice of didactic instruction. Tomasa returned to her classroom and began using activities she had learned from the workshop. Her case raised the concern, however, that the nature of these activities might not be appropriate for teaching third graders. The professional development providers, evidently, recommended the use of alcohol burners; this raised questions concerning the sorts of science experiences that were being promoted as relevant to Filipino children in elementary classrooms.

There are themes across both cases that speak of dilemmas not uncommon to elementary classroom teaching on a global scale. But we wondered whether a historical narrative might help us to understand how the Philippines has apparently evolved toward having some of the legacies that problematize science education reform in many other nations worldwide. The historical narrative that follows highlights some of the major political and educational events that have shaped science education and science teacher education in the Philippines. Ultimately, this overview helps us frame some of the present-day tensions that challenge our current efforts to collaboratively enhance science teacher professional development at WVSU.

A Historical Overview of Science and Education and Science Teacher Education in the Philippines

Tensions represented in the two cases of science teaching and learning previously discussed fit within a larger historical narrative characterized by past political and sociocultural struggles that have shaped education in the Philippines. The philosophy of Filipino higher education today is anchored in an environment of freedom, excellence, and relevance designed to harness, develop, and catalyze the constructive and productive use of the full potentials and capabilities of Filipino men and women (Commission on Higher Education, 1996). The ultimate vision of higher education is that Filipino men and women will become creative, decisive, competitive, and critically thinking and acting individuals who will contribute to:

1. Realization of Filipino identity and strong sense of national pride
2. Cultivation and inculcation of moral and spiritual foundation

3. Attainment of political maturity, economic stability, and equitable social progress
4. Preservation and enrichment of the historical and cultural heritage of the Filipinos as a people and as a nation

Values underlying this vision for Filipino higher education reflect historical legacies of colonialism in the Philippines and present-day competitiveness of the global economic market. The emphasis toward developing a strong sense of Filipino identity says much about geopolitical events that have challenged the lives of citizens in this country. In the section that follows, we present a brief historical overview of how science education and science teacher education was established in the Philippines to contextually understand the professional development of science teachers living in the Panay region.

The Early Days of Science and Science Education in the Philippines

The lives of early Filipinos reflected complex scientific skills, evidenced through their practices of nautical navigation, horticultural and botanical applications, and engineering achievements. The Ifugao rice terraces, for example, are recognized today as one of the world's great wonders because of their engineering mastery. While scientific practices have historically been a part of the indigenous practices of Filipinos, formal science education has more recently taken form amidst the political struggles of the early twentieth century.

Immediately after the American Navy under Commodore George Dewey defeated the Spanish fleet at Sangley Point, Cavite on May 1, 1898, the first American school in the Philippines was established in the Corregidor Islands. This was in accord with the directive of President William McKinley to "fit the people for duties of citizenship." The first teachers were American soldiers who laid down their guns and picked up textbooks almost in the same motion. Eventually, U.S. Army transport ships brought American teachers who were widely dispersed across the island provinces to begin the task of establishing the public school system in the Philippines.

The Filipinos proved to be eager students, and public schools were soon filled to capacity. The increase in enrollment and the number of schools created the need for more teachers. On September 1, 1901, the Manila Normal School (the present day University of the Philippines) was organized to begin training teachers. Eleven American teachers comprised the first teaching staff, and 310 students (of whom only 18 were women) comprised the initial prospective teacher population.

Another highly significant contribution to Philippine education was the Peace Corps movement founded in 1961 by U.S. President John F. Kennedy. Throughout the late 1960s and early 1970s, Peace Corps volunteers arrived to teach in the secondary and tertiary levels of education. At the close of the 1970s, their involvement shifted from classrooms toward involvement in governmental and nongovernmental agencies. Eventually, the Corps' interests shifted back to schools in the 1980s and 1990s—this time to provide teacher training.

Philippine Science Education and Teacher Development through the Decades

Following World War II, leading scientists in the country underscored the woeful state of science teaching from the grade school level to college level. The many problems that beset science education and science teaching in these decades were enumerated by Clark Huber, a science educator of Wheelock College in Boston, Massachusetts, and Fulbright fellow in the Philippines in 1963–1964.

Huber identified language as the foremost problem of science education. A minimum of three languages had to be learned by the children in Tagalog speaking regions, and four to five languages by children in the non-Tagalog speaking areas: Filipino (the national language), English, and Spanish for the former, and all three languages plus the local dialect(s) for the latter. Shortages of resources such as books, teaching aids and devices, and equipment were also cited by Huber as problematic. The ratio of books to students was low, 1 : 4 in Manila schools and only 1 : 10 in the remainder of the country. Whatever books were available were imported publications, so students were more familiar with foreign flora and fauna than Philippine species. Science laboratory equipment was almost nonexistent in schools outside of metro Manila. Access and procurement posed insurmountable difficulties. As a result, teaching was mostly "chalk and talk," and the few laboratory experiments that could be done were large group demonstrations by the teacher.

The 1950s and 1960s saw the first serious attempts to improve science education in the country. A chain of curriculum laboratories was established to strengthen curriculum development efforts. The laboratories were located in the Philippine Normal School (formerly the Manila Normal School) and within eight normal schools located in Iloilo, Cebu, Albay, Ilocos Norte, Pangasinan, Bukidnon, Zamboanga, and Leyte. These sites served as centers for planning, preparation, and production of curriculum materials. In 1957, the Philippine government made the teaching of science compulsory in all elementary and secondary schools.

In the early 1960s, a group of biology educators at the University of the Philippines organized themselves into a team to adapt a U.S.-published textbook, *Biological Sciences Curriculum Study (BSCS), Green Version*. Many other science curriculum resources from local and foreign sources were introduced throughout this decade; most Filipino educators today prefer to adopt publications from foreign publishers and prefer Filipino published texts. During this decade, the Science Teaching Center was established at the University of the Philippines; recently, the center was renamed the Institute for Science and Mathematics Education development (ISMED). Today, ISMED plays a central role in the training of Filipino preservice and inservice teachers in science and mathematics. It should also be mentioned that during this decade, many agencies, including the Peace Corps, assisted in the training of science and mathematics teachers.

Martial law was declared in 1972, and the president authorized major educational reforms. One reform initiative involved the creation of nine regional science-teaching centers (RSTCs). The RSTCs, in addition to ISMED, were responsible for training elementary and secondary teachers and disseminating new approaches and materials to the regions they served. Summer Science Institutes were also organized to cater to the needs of teachers who seriously needed training in both science content and teaching delivery skills. A central focus highlighted in the institutes was development of scientific inquiry skills as a method of teaching/learning and letting students *do* science instead of teaching *about* science. During this time, teacher education programs were gradually revised in teacher training institutions. These revisions allowed new areas of concentration and specialization (including science and math) in the teacher education program for prospective elementary teachers.

The 1980s were characterized as the decade of teacher training to improve the competence of teachers. Training programs were of short-term duration, lasting only several days to a few weeks. The Department of Education, Culture and Sports; UNICEF; and several foundations funded several of these programs. As well, curriculum efforts in textbook development were continued. Curriculum materials development included resource materials for teachers and equipment development that would encourage teachers to use simple apparatuses to teach science.

During the 1990s, the Ford Foundation, UNESCO, the Peace Corps volunteers, and other agencies withdrew from science education projects in the Philippines as funding programs were exhausted or discontinued. A new set of collaborators, however, joined in the pursuit of developing more relevant science education. Japanese volunteers known as JOCV (Japanese Overseas Cooperation Volunteers) were dispatched to the different RSTCs to assist in the development of elementary and secondary science and mathematics education. The Australian government also became involved in science education with the Philippines when it forged an agreement for the training of Filipino elementary science and mathematics teachers at teacher training institutions in Australia. Upon their return, these teachers were expected to conduct inservice training in their service areas. At the same time, a group of Australian elementary science and math specialists assisted the returning teacher-trainers in the Philippines.

In 1998, William Padolina, then Philippines Department of Science Teaching secretary, initiated a plan in science and math education to address the decades-old structural and developmental weaknesses confronting teachers in science and mathematics. This initiative, which is currently underway throughout the Philippines, is referred to as Project RISE: Rescue Initiatives in Science Education. Project Rise is a 3-year project initiated in response to the dismal results of the Philippine participation in the TIMSS, which pointed to the need to look into some aspects of teacher preparation and qualification. The training program is aimed at increasing the competence as well as the confidence of science and mathematics teachers in the elementary and secondary levels in public and private schools of the country to teach their subject areas effectively. The training sessions occur during school days and on weekends at classes offered through the RSTCs. It is expected that by year 2003, pupils and students of teachers trained through Project Rise will have gained the competencies to become comparable with pupils of other countries in science and mathematics.

Changing Traditions in the Professional Development of Science Teachers

The historical overview of professional science teacher development, and science education in general, presents a long tradition of outsiders playing significant roles in the educational and political arenas of the Philippines. A result of this tradition is that professional development and curriculum design are relegated to external agents instead of local practitioners. This is problematic in two regards. First, it does not recognize the knowledge Filipino educators themselves bring to science education, and second, it risks decontextualizing science teaching and learning. Consequently, a primary interest we shared as a research team was to involve teachers as researchers of their own teaching practices. We also wanted to explore what it might mean to contextualize science teaching and learning with respect to the cultural landscape of Iloilo. Finally, given that much of the science curriculum and teacher training had seemingly reflected a secondary science orientation, we hoped to encourage the emergence of elementary science pedagogy. In this section, we reflect on our experiences and what we learned with respect to interests undergirding our collaborative work in the Philippines.

Involving Teachers as Researchers

The history of science teachers' professional development in the Philippines resembles teacher training worldwide, wherein teachers are assumed to be consumers of knowledge. The history and experience of teachers are not seen as relevant to their practice; rather, teaching is viewed as merely a technical practice of applying generalized teaching strategies (Bullough & Gitlin, 1991). More contemporary approaches to science teacher education have adopted the idea of teachers as "reflective practitioners"—a perspective that regards teachers' thinking in and on their actions as centrally important to their classroom practices (e.g., Grimmett & Erickson, 1988; Schon, 1983; Zeichner & Liston, 1987). Reflective practice is essential to action research, wherein teachers become inquirers into their own practices. Action research has at its heart a "democratic epistemology" (Noffke & Brennan, 1991, p. 187) that recognizes that knowledge is the currency of power and change; thus change in education calls for teachers to generate knowledge they believe is necessary to enhance their practices in the classroom. Reflective practice and action research are not intended to be practiced in isolation; rather, when practiced within a community of practitioners these methods give teachers access to alternative ways of interpreting and evaluating practices, which they can juxtapose with their own assumed views.

Science teacher educators at WVSU have made recent shifts to reflect their interest in using action research as the basis for teacher learning in their science teacher education program. Prospective teachers are expected to conduct community-based action research as part of their student teaching experience. One example of an action research project conducted by a student teacher involved cleaning up school grounds at a rural elementary school. This particular school is located in a "squatters" section of Iloilo, where it is not uncommon to find families living in shelters made of cardboard. The school had suffered serious neglect under the direction of a former administrator. The school had received a new principal at the start of the school year who was eager to revitalize the school and the surrounding community. The student teacher, inspired by the principal, organized parents and students to work after school and on weekends to clear away vegetation, which had almost completely overgrown the entire school. The student teacher's collaborative project helped encourage parents to enroll their children in school and to feel a sense of community ownership toward the school. Additionally, science teacher educators are now looking closely at ways for using action research to generate science understandings through the projects.

Another shift from traditional professional development took place through our collaborative research activities. In our research project, teachers were invited to write a case that reflected a dilemma they had experienced in their science teaching. Various dilemmas were identified as teachers presented and discussed their cases. One dilemma—elementary teachers' reliance on textbooks for science knowledge—was illustrated in the case written by Rossini G. Monsalud (Case #1). The issue of elementary teachers feeling they lacked knowledge to teach science was reiterated in several other teachers' cases as well. Discussions about this dilemma encouraged participants to explore their assumptions about science being viewed as facts to be taught accurately to students versus science being practiced as inquiry in the classroom. Conversations wove in and around a variety of teachers' perspectives but never yielded a single "correct" interpretation of how the teachers should think.

As participants listened to each other explore possibilities for thinking and practicing science teaching, there was a growing sense that teaching involves working through uncertainties and that there are multiple ways teachers might try to resolve their dilemmas. This was an eye-opening insight as participants in the case experience indicated they had not previously recognized the problem-solving aspect of teaching practice. Shifts took place as teachers began to see each other as important sources of knowledge for their teacher learning and to recognize that science teaching is a complex activity for which, often, there are no single or simple solutions. One experienced teacher, Mariet, indicated she had not previously realized that prospective teachers are not merely trained and that they experience struggles in learning to teach: "Before this activity we did not know that student teachers also have dilemmas. We thought they are just following what we are doing and what the book says. Before, my student teacher was quite hesitant to ask or tell me anything. But after the discussions we had, she became open to me."

Participants overwhelmingly agreed that reflective practice, such as using case-based pedagogy, should be used to support their professional development. Another teacher commented:

> The discussion part has really brought me some insights. . . . We talked of what's inside us . . . and we discovered that student teachers are also persons . . . sensitive and could be embarrassed. . . . This gives us an insight that we should be careful in dealing with them because they too have feelings. . . . Whether you are a teacher, a cooperating teacher, a student teacher . . . all ideas are accepted, no one is greater or lower . . . no one is higher . . . nobody is lower . . . everybody is accepted. Everybody feels happy to be in the group and to be discussing without restrictions. You tell your experiences and everything is accepted and after the session you are enlightened and your heart is lighter. . . . With my student teachers . . . we are more open to each other now.

There was a sense that study participants not only valued knowledge they generated through the case discussions, but also that the experience encouraged them to become colearners in the classroom. The idea of creating teacher-learning communities as the generative site for developing knowledge for science teaching represented a shift away from the expecting outside experts to provide the knowledge needed for teaching.

Balancing Indigenization and Globalization

Our collaborative experience raised issues for the research team to ponder with regard to rethinking the science teacher education curriculum. The case written by Tomasa Ferando (Case #2) reflected a common problem of modeling the use of appropriate science teaching practices for elementary science learning. Tomasa was encouraged to use resources for teaching science that were unsafe for use among young children. Furthermore, visits to schools revealed gardens being used to teach children about indigenous plants and ecological practices, but the science centrums (specialized rooms for science instruction) displayed artifacts such as microscopes, encyclopedias, and periodic tables to represent science. There are more relevant objects teachers could use to support inquiry with their children, but teachers are encouraged to use objects that reflect a Western-based science history. Such representations of science are troubling not only because they suggest science must employ the use of formal science equipment, but also because they reflect an underlying European or Western history of science. Tomasa's case conveyed deeper tensions associated with indigenous knowledge versus the rhetoric of global scientific literacy (see Arellano et al., 2000a). Filipino teachers indicated that they are often using textbooks written by non-Filipino publishers. Problems arise when information about Filipino objects (e.g., plant or animal species) is not accurately represented or examples presented in texts are not found in the Philippines. The issue of indigenous knowledge versus global scientific literacy has become a focal point that the research team is continuing to explore. The research team is beginning to explore possible resolutions to this dilemma by finding examples of indigenous knowledge and practices that can be used to embed Filipino science teaching and learning within local cultural and historical situations The case-based experience models what it means to develop a more contextually relevant science curriculum and science teacher education.

Our case experience indicated that tensions have been created as non-Filipino and nonelementary persons and curricular resources have been predominantly used for training elementary Filipino teachers to teach science. Through the case experience we began to hear possible beginning points for engaging prospective and practicing elementary teachers in learning to teach science. Tomasa's case, for example, suggests that we should explore with teachers materials they will find more inviting to use in conducting science inquiry with children. Also, it will be important to reflect with teachers on their views of science so that instead of viewing themselves as being deficient in their knowledge of science they might be encouraged to adopt a stance of being co-inquirers alongside their students as the basis for generating scientific ideas.

Looking to the Future: Trends and Issues in Filipino Science Teacher Professional Development

Science teacher professional development in the Philippines is situated between a historical legacy of international involvement in science education reform and present-day concerns for preserving cultural identity. Curricular tensions in science education reform are emerging between concerns to promote indigenous curricular interests and the improvement of Filipino performance in global science competitions. Beyond the layer of rhetoric of policies and evaluations, we listened to prospective and practicing teachers. Through our collaborative research we were reminded of the complexity of daily experiences and perspectives that teachers and their students bring to the classroom landscape. While it is important for nations such as the Philippines to establish standards for science education reform, there is the need to ensure that science education contributes in ways that help teachers and students contribute to the well-being of their local communities. In our concluding comments, we want to consider what sorts of legacies might be important to consider with regard to Filipino science education reform and alternative ways for envisioning new approaches to support science teacher professional development in the Philippines.

Legacies for the Future

The people of the Philippines have experienced a long history of non-Filipino influence on their science education curriculum and teacher training. External influences on their science curricular developments and professional development programs continue to mediate science education reform in the Philippines. Filipinos have been greatly challenged in terms of building up their own agency and resources to support their own science teacher professional development and adoption of Filipino-created curricular materials. The geographical character of the Philippines as well as political unrest—which continues even today—makes it difficult to systematically communicate and physically support goals for science education reform across this nation. These conditions have created a legacy of Filipinos needing to develop a sense of identity and purpose toward creating visions of science education reform by and for the people of the Philippines.

Through our collaborative research experience at WVSU, it appears that another legacy of professional development practices in the Philippines has reflected the use of "outside experts" as providers of knowledge and training to Filipino teachers. This has created problems as science education curricula represent decontextualized science and Filipino knowledge of science teaching has become devalued. The use of action research poses the possibilities for gathering insights from people in communities about their science experiences and what forms of science knowledge they might value. This information could be shared among teacher-learning communities—not unlike the group represented in our case experience research. Localized scholarship would validate the contextualized science curriculum this approach would generate and would assist teachers in developing pedagogical practices appropriate to their local communities.

Finally, the rhetoric of globalization in science education needs to be critically examined for underlying ideologies implied through these discourses. In recent literature, for example, there have been discussions about developing globally scientific literate citizens (Kyle, 1999; Lee, 1997). One has to consider whose knowledge and what scientific issues will be deemed important—presumably these would be issues of importance to persons living in First World nations. International comparisons of mathematics and science teaching and learning through evaluations such as the TIMSS reinforce messages about what sorts of knowledge and educational practices are most valued. Conversations with members of the research team, as well as among other science educators around the country, reveal concerns about the reported poor performance of Filipino students in mathematics and science on the TIMSS. However, there is also recognition that great strides have been made toward dealing with linguistic diversity in classrooms and in curricular resources and toward developing an appreciation to identify and preserve indigenous knowledge. When Filipino educators examine results on the TIMSS retest, they will need to interpret those results within an interpretive framework contextualized within their nation's science teacher education reform. Current efforts to recognize and deal with cultural and language diversity will help Filipinos appreciate their unique identity and the educational challenges that come with trying to preserve their cultural heritage in light of science curricular changes taking place.

Potentials for Future Collaboration

As we consider the interplay of past and future legacies in changing science teachers' professional development in the Philippines, it encourages us to explore the potentials for future collaborations involving international colleagues. There is concern for breaking the traditional notion that outsiders bring expert knowledge needed to change Filipino teaching. Changing the language used to describe the roles assumed among collaborators can be an effective way to shift away from the expert model toward a model of collegiality. In our collaboration, we adopted the metaphor of "teacher community" to reflect our shared interest to learn with and from each other.

We also wanted to shift away from formalistic research traditions that have tended to privilege the assumptions of the researcher and have distanced the researcher from those being researched. Action research was an alternative research perspective that had interested science teacher education colleagues because of its emphasis on democratizing the purposes and approaches used in conducting research. By chance, our research team was drawn together as we recognized our mutual interest for the underlying tenets of action research. We used case-based pedagogy as a vehicle for engaging study participants and our research team in dialogues about science teaching. Ultimately, we saw the community-building experience that supported our learning create a fundamental shift in our views of professional development practice and research. This is a shift away from teacher workshops, which have traditionally used predetermined goals and objectives and emphasized the dissemination of specified knowledge and strategies for science teaching. Our experience moves beyond this to see that when teachers see themselves as generators of, and contributors to, knowledge in a teacher learning community they develop a sense of agency for their own professional development. At the same time, we experienced a shift in our research as we began to look beyond the texts of data for meaning to see contexts of experiences as the primary landscape of our learning. Our potentials for future collaboration have been enhanced as we have a better understanding of each other's needs and goals for learning. It is our hope that by sharing our current collaborative research others will be encouraged to find new ways to participate in building local and international science teacher learning communities.

References

Abell, S., Dennamo, K.S., Anderson, M.A., Bryan, L.A., Campbell, L.M., & Hub, J.W. (1996). Integrated media classroom cases in elementary science teacher education. *Journal of Computers in Mathematics and Science Teaching, 15,* 137–151.

Arellano, E.L., Barcenal, T.L., Bilbao, P.P., Castellano, M.A., Nichols, S.E., & Tippins, D.J. (2001a). Case-based pedagogy as a context for collaborative inquiry in the Philippines. *Journal of Research in Science Teaching, 38*(4), 1–27.

Arellano, E., Barcenal, T., Bilbao, P., Castellano, M., Nichols, S., & Tippins, D. (2001b). Using case-based pedagogy in the Philippines: A narrative inquiry. *Research in Science Education, 38*(5), pp 502–528

Beaton, A.E., Martin, M.O., Mullis, I.V.S., Gonzalez, E.J., Smith, T.A., & Kelly, D.L. (1996). *Science achievement in the middle school years: IEA's third international mathematics and science study.* Chestnut Hill, MA: Center for the Study of Testing, Evaluation, and Educational Policy Boston College.

Beaton, A.E., Mullis, I.V.S., Martin, M.O., Gonzalez, E.J., Kelly, D.L. & Smith, T.A. (1996). *Mathematics achievement in the middle school years: IEA's third international mathematics and science study*. Chestnut Hill, MA: Center for the Study of Testing, Evaluation, and Educational Policy Boston College.

Bracey, G.W. (2000). The TIMSS "final year" study and report: A critique. *Educational Researcher, 29*(4), 4–10.

Bruner, J. (1986). *Actual minds, possible worlds*. Cambridge, MA: Harvard University Press.

Bullough, R.V., & Gitlin, A.D. (1991). Educative communities and the development of the reflective practitioner. In R. Tabachnich & K. Zeichner (Eds.), *Issues and practices in inquiry-oriented teacher education* (pp. 35–55). New York: The Falmer Press.

Clandinin, D.J., & Connelly, F.M. (1996). Teachers' professional knowledge landscapes: Teacher stories, stories of teachers—school stories, stories of schools. *Educational Researcher, 25*(3), 24–30.

Commission on Higher Education. (1996). *Long term higher education development plan (1996–2005)*. Manila, Philippines: Commission on Higher Education.

Connelly, E.M., & Clandinin, D.J. (2000). *Narrative inquiry: Experience and story in qualitative research*. San Francisco, CA: Jossey-Bass Publishers.

Grimmett, P., & Erickson, G. (1988). *Reflection in teacher education*. New York: Teachers College Press.

Hernandez, D. (1996). *History, philosophy and science education*. Monograph no. 49. Quezon City, Philippines: University of Philippines Institute of Science and Mathematics Education.

Howe, A., & Nichols, S. (2001). *Listening to teachers: Case studies in elementary science teaching*. Upper Saddle River, NJ: Merrill/Prentice Hall.

Kagan, D. (1993). Contexts for the use of classroom cases. *American Educational Research Journal, 30*, 703–723.

Koballa, T., & Tippins, D. (2000). *The promise and dilemmas of middle and secondary science teaching: A classroom case handbook*. Upper Saddle River, NJ: Merrill/Prentice Hall.

Kyle, William C., Jr. (1999). Science education in developing countries: Challenging first world hegemony in a global context. *Journal of Research in Science Teaching, 36*(3), 255–260.

Lee, O. (1997). Scientific literacy for all: What is it, and how can we achieve it? *Journal of Research in Science Teaching, 34*(3), 219–222.

Loucks-Horsley, S., Hewson, P.W., Love, N., & Stiles, K.E. (1998). *Designing professional development for teachers of science and mathematics*. Thousand Oaks, CA: Corwin Press.

Noffke, S.E., & Brennan, M. (1991). Student teachers use action research: Issues and examples. In R. Tabachnich & K. Zeichner (Eds.), *Issues and practices in inquiry-oriented teacher education*, (pp. 186–201). New York: The Falmer Press.

Polkinghorne, D.E. (1988). *Narrative knowing and the human sciences.* Albany: State University of New York Press.

Schon, D. (1983). *The reflective practitioner.* New York: Basic Books.

Zeichner, K., & Liston, D. (1987). Teaching student teachers to reflect. *Harvard Educational Review, 57,* 23–48.

Chapter 6
Science Teacher Education in Trinidad and Tobago
Challenges and Possibilities

JUNE GEORGE

Teacher Education in the Commonwealth Caribbean

The Commonwealth Caribbean consists of those Caribbean territories that were formerly under British rule. The education systems in these territories are fairly similar since they were all patterned after the British system in the colonial days. Even though most of these territories are now independent, vestiges of the British system persist and the structures for teacher education have been slow in changing. Typically, training opportunities for primary school teachers are provided at a teachers' training college and those for secondary school teachers are provided at the regional University of the West Indies (UWI). Within recent times, however, the teachers' colleges in some territories have taken on the added responsibility of preparing secondary school teachers. The preservice teacher education model exists in only a few of the territories. More often than not, individuals enter the service as untrained teachers, teach for a few years (the length of time varies from country to country), and then enter the teacher training institution for their professional training.

The Trinidad and Tobago Context

Trinidad and Tobago is a twin-island republic situated at the end of the chain of Caribbean islands. Like many former British colonies, its education system has been modelled along the lines of the British system. The two main sectors in the public school system are the primary sector (of 7–9 years' duration) and the secondary/high school sector (of 5–7 years' duration).

Although primary education is free and compulsory, education beyond this point is not universal. This is because of limited space in the secondary sector. Selection for placement in the secondary level is based on performance on the national Common Entrance Examination (CEE), which is taken after 7 years of primary schooling. Students who do not secure a place in the secondary sector may continue for a further 2 years at the primary level and then leave for the world of work or pursue vocational education at youth camps or youth centers. Students in the secondary sector sit terminal examinations set by the regional Caribbean Examinations Council (CXC) at the end of 5 years of high school. Many students leave high school at this point to join the labor market or to pursue further studies at nonuniversity tertiary institutions. Those students who have performed well at the CXC examinations continue in high school for a further 2 years and then sit the University of Cambridge General Certificate in Education Advanced Level examinations, which are qualifying examinations for entry into university.

Science is a compulsory subject for all students at the primary level. Science: A Process Approach for Trinidad and Tobago (SAPATT) is the official science curriculum for primary schools in Trinidad and Tobago. It was patterned after the American SAPA—Science A Process Approach. The curriculum is very heavily biased toward the science processes; the science content/concepts in each lesson serve only as the vehicle for teaching/learning the designated science processes. Primary science teachers (who are not university graduates) have found this approach to science teaching extremely challenging.

At the high school level, science teachers are usually science graduates with a bachelor of science degree. Science has enjoyed a privileged position on the high school curriculum over the years, since it is thought to be a difficult subject that is within the capabilities of only the more academically inclined students. This position is changing now because of two main factors. First, the harsh economic realities of the mid 1980s and 1990s have generated a marked interest in the area of business and in business studies in the schools. It is not uncommon now to find very academically gifted students at the upper high school level specializing in business studies in preference to science. Second, the global call for "science for all" has found a niche in these islands, to the extent that many high schools now encourage their students to study at least one science subject right through to the CXC examinations at the end of 5 years of high school. This opening up of the study of science to the general high school population, regardless of academic ability, has posed new challenges for science teachers at this level, most of whom are university graduates.

Provisions for the Formal Training of Science Teachers

Emerging from the British tradition, primary school teachers are not normally university graduates. Typically, these teachers would have graduated from high school with satisfactory performances in at least five subjects at the CXC level, including a science subject. After having taught in a primary school for 2–3 years as untrained teachers, they would then have attended a 2-year primary teacher-training program at one of the two government-run teachers' training colleges.

At the teachers' colleges, a compulsory core curriculum covers the general education foundation areas, courses on teaching methodology, and content courses in the various subjects taught at the primary school. The science content curriculum at the colleges has been designed to mirror the process-based curriculum in use in the primary schools. Teacher trainees at the teachers' colleges must also choose one subject (called an elective) in which to specialize at a more advanced level. Typically, the advanced science group consists of no more than 15 students at each college. This is but a small percentage of the yearly group population, which is usually about 200 students at each college.

The School of Education at the local campus of the UWI conducts formal teacher training beyond the teachers' college level. Graduates of the teachers' colleges may seek professional advancement through the Certificate in Education (Cert.Ed.) and Bachelor of Education (B.Ed.) programs offered by the School of Education of the local campus of the UWI. The Cert.Ed. in the teaching of science is one of the certificate programs offered. It is a 300-hour program that is accessible to teachers who have been successful in the program offered by an approved teachers' college. It consists of courses in Foundations of Education, Measurement and Research Methods, The School as an Organization, The Practice of Education, Concepts in Science, and The Teaching of Integrated Science.

Holders of the certificate in education may spend a further two years in pursuit of a B.Ed. degree. To obtain the degree, such candidates must obtain a further 60 credits, which must include further work in Foundations of Education, cross-faculty courses, professional specialization courses, and a practicum. Although there is no specific B.Ed. degree in the teaching of science, candidates may choose two professional specialization courses that focus on the teaching of science. They may also locate their work for the practicum in science and may pursue cross-faculty science content courses in the Faculty of Agriculture and Natural Sciences.

The Diploma in Education (Dip.Ed.) program, which is a 1-year inservice program, is the avenue for the professional development of science teachers (and other teachers) at the secondary level who are holders of a bachelor's degree. Teachers pursuing the Dip.Ed. would typically have taught in the high school for 4–5 years before gaining entry to the program. This is sometimes a choice on the teacher's part, since a professional qualification is not required for employment as a high school teacher. It may also be because there is a large backlog of untrained high school teachers and younger teachers must wait their turn in the selection process. About 100 teachers are trained in the Dip.Ed. program each year. Teachers continue to teach in their respective schools while in the program. They attend lectures during school vacations, beginning with an intensive session in the summer. They also attend lectures for one full day during the week during term time. University lecturers visit the teachers in their home schools at intervals to monitor and tutor them in their classroom practice.

The Dip.Ed. in the teaching of science provides courses in Foundations of Education and a specialization in the teaching of science. About 30 high school science teachers are trained in this program each year. The course is taught as a course in the teaching of science generally, but attempts are made to assign students who teach mainly one of the separate science subjects to a tutor who has expertise in that area. There are no programs at the master's level that offer a specialization in the teaching of science. At the M.Ed. level, students who are pursuing the teacher education specialization may opt to take a module called Science in the School Curriculum. This is the only science education course offered at the master's level. Students may, however, pursue research projects and theses in the area of science education. Whereas the program at the teachers' colleges is offered on a full-time basis, those at the School of Education are offered on both a full-time and a part-time basis. The reality, however, is that most of the students at the School of Education are practicing teachers who attend classes on a part-time basis on evenings. Teacher trainees at the teachers' colleges continue to receive their full salary from the government while undergoing full-time training.

Orientation in Science Teacher Education Programs

The Cert.Ed., B.Ed., and Dip.Ed. programs all have a similar orientation. The underlining philosophy is that students, regardless of their area of specialization, should study foundation areas such as the philosophy, psychology, and sociology of education; language in education; assessment and review; and research methods. They should also attempt to draw on these foundation areas in their work in the classroom. All of the programs require teachers to engage in a field/classroom investigation of an educational problem. The aims of the science components of all of the programs are similar, with the depth of coverage increasing from the Cert.Ed. level to the Dip.Ed. level. Emphasis is placed on the nature of the discipline, learning theories in science, science education in relation to other aspects of the school curriculum, designing and selecting learning activities in science, and laboratory and classroom management. Teaching practice is also an important component of the Cert.Ed. and Dip. Ed. programs. This is conducted through school-based tuition by university personnel and is assessed using performance assessment strategies.

Other Avenues for the Professional Development of Science Teachers

The Ministry of Education mounts workshops periodically for both primary and secondary science teachers. Sometimes, these workshops are a joint effort of the Ministry of Education, the Association for Science Education of Trinidad and Tobago (the local science teachers' association), and the School of Education of the UWI. Issues that are of specific concern to science teachers are addressed at these meetings.

The regional CXC also assists in the professional development of science teachers. Although its main function is to set and administer examinations in the various high school subjects for students of the contributing Caribbean territories, CXC has found it necessary to widen its scope of activities to include some teacher training as well. It does this through periodic workshops in areas that teachers find to be challenging. With respect to the teaching of science, CXC has mounted workshops to train science teachers to deal with the innovation of school-based assessment (SBA) of practical work in science. The SBA is designed to facilitate teachers' assessment of each student on a set of practical skills and also on his or her disposition to practical work in science. The skills tested include observing, recording, reporting, interpreting data, planning and design of experiments, drawing, and environmental awareness. This performance-based approach to assessment in science is new for many teachers, and CXC found it necessary to help alleviate the problem by providing training for such science teachers.

The Association for Science Education of Trinidad and Tobago also assists in the professional development of science teachers through its publications. A yearly journal and a magazine published two or three times in the school year help to serve this function. Practicing teachers are encouraged to submit articles for consideration for publication so that innovations/problems/issues in science education can be shared and discussed within the local science education community.

Some Challenging Issues

The efforts described here result in the production of trained science teachers for both levels of the system. Yet problems persist. These problems relate to the confidence level of primary science teachers, the numbers of trained high school science teachers produced annually, support systems for teaching science in the schools, and the relevance of the science taught.

Confidence Level of Primary School Science Teachers

As indicated earlier, primary school teachers are mainly high school graduates who have been trained at the teachers' colleges. In most primary schools, one teacher teaches all the subjects to a particular class. Most primary school teachers, therefore, teach science. Many of these teachers have a weak background in science. Even though they had met the entry requirement to the teachers' college of a satisfactory performance in one CXC science subject, there are gaps in their science content knowledge (particularly in the physical sciences), despite their exposure to a science content course at the teachers' college. To compound the problem further, many of these teachers are exposed to the process approach to teaching science for the first time at the teachers' college. The upshot of all of this is that many primary science teachers lack confidence in teaching science and often resort to teaching strategies that involve little laboratory work and that present the process approach in the form of lesson notes. A few schools have experimented with using a teacher with a strong science background as the specialist science teacher (see, for example, Edwards, 1991) with some degree of success. However, it is difficult to sustain such efforts because of staffing quotas and staff shortages.

Supply of Trained High School Science Teachers

With over 100 high schools in the country, the supply of approximately 30 trained high school science teachers is woefully inadequate. Added to this is the fact that, because science graduates command better salaries in the private sector than in teaching, some high school science teachers (including those who have professional training) switch to the private sector when the opportunity arises. This means that teachers who have had no professional training teach a fair proportion of high school science students.

Support Systems for Teaching Science

In a survey of lower secondary science teachers (George, 1995), respondents cited the lack of proper resources for teaching science as one of the main obstacles in their work. Teachers cited the lack of proper equipment and resource materials as serious handicaps. At the primary level, the situation is perhaps worse, since many of the schools do not have science laboratories or science rooms. It should be noted, however, that some teachers are able to rise above these shortcomings and engage in creative improvisations in conducting their science classes.

The Relevance of the Science Taught

The training programs available to teachers of science in Trinidad and Tobago attempt to deal with the nature of science, the theories that attempt to illuminate how children learn science, and the strategies that might be employed for teaching science. These are worthwhile components of any program on science teacher education. What is missing from the local programs, however, is a concerted effort to situate the teaching/learning of science in some of the experiences of students that are unique to the local setting. While there are attempts to focus on the relevance of science to the lives of the children, the concept of relevance usually employed is a restricted one that embodies only those aspects of students' lives that map readily on to conventional science. For example, a unit of work on heat would typically deal with conduction, convection, and radiation, with examples of these processes drawn from the local setting. It is hardly ever the case that the training of the science teachers (and the subsequent classroom teaching) would also focus on different conceptions of heat that might be unique to the local context. In the local setting (and in some other settings as well), the human body is thought to become "heated" when exposed to high temperatures and after the ingestion of certain foods (George, 1995). These conceptions of heat are different to those espoused by conventional science, but, they are relevant to the lives of the students. The science taught can only be called relevant if these local conceptions are examined in the science classes. Teacher training programs should prepare teachers to deal with such situations, but this is hardly being done at the moment.

Some Recent Innovations

As the more traditional approaches to the professional development of teachers continue, a few initiatives are breaking new ground. These initiatives have been occurring outside of the formal science teacher education structures.

In March 1997, a project was mounted at the School of Education, UWI, Trinidad, to train high school science teachers to produce science teaching resource materials that are contextualized and relevant to the lives of the students. The project had the full support of the Ministry of Education and received some funding from United Nations Educational Scientific and Cultural Organization. Nineteen high school science teachers from throughout the country were engaged in this pilot project. These teachers were chosen because of their previous engagement in science curriculum development efforts organized by the Ministry of Education.

The project was initiated through two 4-day workshops. The overall goal of these workshops was to expose the science teachers to the procedures for designing contextualized science teaching materials and to begin the production of such materials. The specific strategy for producing contextualized materials was adapted from methods used by a consultant to the project (Lubben) and his coworkers in a similar situation in Swaziland (Lubben, Campbell, & Dlamini, 1996).

The teachers were first familiarized with contextualized materials. They were then asked to detail contexts pertaining to a specific unit on the official science curriculum for the lower secondary level, which they thought would be relevant to the lives of youngsters in Trinidad and Tobago. A lesson format was agreed upon. Each lesson begins with stimulus material depicting some aspect of everyday life that is within the experiences of students. Students would then be asked to suggest explanations for the situation with their current understandings of science. A line of inquiry/investigation would then be developed in which the related science concepts/principles are introduced. Students would then be required to apply their new understandings to the situation that was considered at the beginning of the lesson.

Teachers worked on the lesson outlines in small groups, and whole group discussions were held at intervals to allow for the exchange of ideas. In all, 33 lesson outlines were produced—18 on the unit Maintaining Health and 15 on the unit Water: An Essential Resource. The lesson outlines were subjected to preliminary editing before being piloted in the schools.

The workshop teachers were very enthusiastic about the project. Through the exercise, they devised additional objectives for the official curriculum document, which they considered to be essential if the science taught is to be relevant to the students' lives. They included materials that they would not normally use in their teaching but which they came to realize are truly part of students' everyday experiences. For example, in the unit on water, they included lessons on the reasons why some people do not receive tap water and the efficacy of a truck-borne water supply system. In the unit on maintaining health, they designed lessons on the impact of excessive noise from discotheque music and the music of the steelband (the national instrument) on one's hearing.

This method of professional development of science teachers has had both benefits and drawbacks. The benefits have been many. Teachers enjoyed the opportunity to be able to meet with colleagues from different schools to discuss their work in an inviting atmosphere. More importantly, however, teachers, through their own words, welcomed the opportunity to review their approach to science teaching.

Inherent in the use of the contextualized approach were some benefits. Teachers were forced to think seriously about the *real* everyday experiences of their students (as opposed to the "sanitized" ones that are often presented in textbooks). The focus on students' everyday experiences led them to seek resources from sources that they would not readily have tapped (e.g., the local market, the nightly television weather forecast, the local newspapers). They were also challenged to be creative. One teacher confessed that she had always thought that she lacked creativity but that she had had second thoughts after the workshop experience!

But there were also some challenges. The first challenge that teachers had to face was that of thinking of the students' everyday experiences first in the planning of a lesson. Teachers' natural inclination was to identify the science to be taught as the first step in planning a lesson. The strategy of focusing first on the students' everyday experiences was only mastered after considerable practice.

Another major hurdle encountered was that of analyzing the everyday experiences to see what were the links (if any) with conventional science. There were instances where no discernible links could be identified. For example, in a brainstorming session to identify the everyday experiences of students, teachers had indicated that some students take herbal medicines for conditions such as asthma and menstrual cramps. The practice of using herbal medicines for certain ailments is fairly widespread, particularly in rural areas. Yet the chemical composition of many of these herbs and their likely effects on the human body remain largely unresearched by conventional scientists. Teachers were handicapped by this lack of knowledge, and such everyday experiences were ignored in the material writing exercise.

Despite the challenges, however, these science teachers rated the workshop highly, and they indicated that they had been rejuvenated to return to the classroom to try out these materials.

Another recent innovation involved primary school teachers, not only from Trinidad and Tobago, but also from the rest of the Commonwealth Caribbean. This regional workshop, called Science and Mathematics Integrated Learning Experience, was sponsored by the Caribbean Council for Science and Technology and funded by the Organization of American States. It was held on the island of St. Thomas in the U.S. Virgin Islands in August 1998.

The participants of this workshop were two primary school teachers, a principal, a science curriculum officer, a mathematics curriculum officer, a science teacher educator, and a mathematics teacher educator from each of the various territories. The mixed group, with participants from the various segments of the primary education sector, was chosen in an effort to minimize problems of implementation when participants returned to home base. The three main goals for the workshop were:

- To demonstrate creative approaches to using the environment for teaching science and mathematics
- To develop strategies for using the environment to implement an integrated approach to teaching primary science and mathematics in countries of the Caribbean
- To expand the resource base available to Caribbean teachers and trainers in the development of curricula and training programs.

The workshop was conducted at two sites—the University of the Virgin Islands and the Virgin Islands Environmental Resource Station, the latter being a campsite. Participants were divided into groups and the groups were rotated so that each one spent a 5-day period at each of the two sites involved in workshop activities. The instructional strategy used was that of learning by doing. Thus, participants were actively engaged in all the activities and some time was spent discussing how these activities might be used in the primary science classroom.

The activities were pertinent to a wide range of topics/issues that form part of the primary school science and mathematics curriculum. They included the use of the World Wide Web as a teaching resource for primary mathematics and science, computers in teaching primary mathematics and science, the use of calculators in the teaching/learning of primary mathematics, science activities with a hands-on/experimental focus, mangrove ecosystem mapping, comparing coastal and inland ecosystems, scavenger hunt in the natural environment, and examining the characteristics of wave motion at the seaside.

The author served as the facilitator for one of the science activity sessions. This session involved having participants design activities using materials readily found in the home to explore the concepts associated with solutions and suspensions. The objectives for this workshop were:

- To promote the use of the home environment to teach science
- To expose participants to a learner-centered science activity using materials from the home
- To have participants revisit the science concepts and principles inherent in the activities conducted
- To make use of skills possessed by most primary teachers (e.g., chart making)
- To have participants examine the science activity from a primary teacher's perspective by having them propose how it might be modified for use at the various levels of the primary school system

Participants created outlines for several lessons for different age groups. The activities were tried out and outlines revised based on the results of the hands-on experiences. The skill of planning and designing an investigation featured prominently in these activities. The reworked lesson outlines were compiled into a unit with the title Mixing It Up.

Generally, the objectives for the sessions were achieved. Participants' skills in chart making were very evident in this part of the workshop. However, there were instances where misconceptions about some basic science concepts surfaced.

Discussion

The two innovations described in the last section involved small groups of science teachers in projects funded by external agencies. These are efforts that fall outside of the mainstream science teacher education programs run by the Ministry of Education and the UWI. Because these innovations have been externally funded, money has not been much of a constraint and participants have enjoyed very comfortable work settings and access to a wide range of resources. Science teachers have been highly motivated by these experiences.

But the strength of these efforts is also their weakness. Projects such as these are more often than not short-lived and are conducted in environments that do not reflect the reality of the classroom situation. The ideal conditions under which instructional materials are developed and new strategies are tried out simply do not exist in the average classroom. Consequently, after the halo effect has rubbed off, the teacher must face reality and adapt to the local conditions. In addition, in the absence of the support of workshop facilitators and other workshop participants, teachers must "face the music" on their own. It is at this point that the fallout effect can be great, with some teachers reverting to business as usual. The two projects described here were short-lived. Funding came to an end and teachers had to draw on their inner reserves and their creativity to keep the ideas alive in the classroom. There has been some fallout.

Perhaps the solution lies in situating these professional development efforts within the school districts rather than at a central point. This would place an added burden on the organizers as several different workshops would have to be organized and, unless the Ministry of Education begins to include such activities in its budget, a larger amount of external funding would have to be sought. Yet the long-term effect might be greater since more science teachers in a given district would be exposed to innovation and the possibilities of networking science teachers (and other personnel) in the local community after the workshop sessions conclude. These workshops may also serve to provide some measure of preservice training for some of the untrained science teachers in the system.

In addition to the logistical issues described here, the question remains of what the content and focus of these workshops should be. I see these non-mainstream science teacher education efforts as serving to build on the traditional science teacher education programs provided by the institutions by tackling issues and problems that may not be easily tackled within the confines of an academic program, with its need for assessment and other constraints. As indicated earlier, the issue of a truly relevant science curriculum for Trinidad and Tobago is a burning one and should be explored more fully from a cultural perspective (see, e.g., George, 1999). Also, the whole business of alternative forms of assessment as they pertain to learning in science needs to be explored. Further, the newer technologies are beginning to have an impact, even in the rural areas. How do children marry these technologies with their cultural practices and beliefs, and how can science teachers plan for these complex situations in their science classes? These and other issues would need to be addressed if the future education of our science teachers is to be relevant.

References

Edwards, A. (1991). *A case study of the implementation of the SAPATT curriculum in Trinidad and Tobago*. Unpublished master's research report, The University of the West Indies, St. Augustine, Trinidad and Tobago.

George, J. (1995). *The needs, practices and beliefs of lower secondary science teachers in Trinidadand Tobago: A research report*. Trinidad and Tobago: The University of the West Indies.

George, J. (1999). Worldview analysis of knowledge in a rural village: Implications for science education, *Science Education, 83*(1), 77–95.

Lubben, F., Campbell, B., & Dlamini, B. (1996). Contextualised science teaching in Swaziland: Some student reactions. *International Journal of Science Education, 18*(3), 311–320.

Chapter 7
Theory and Practice in Science Teacher Education
The German Experience

HELMUT FISCHLER

Teacher education in Germany has undergone some fundamental changes in recent decades. Some changes were driven by structural changes within the school system that made it necessary to adopt new teacher education regulations. Others were influenced by research findings that prompted political decision-makers to alter aspects of the teacher education system to improve teaching and learning. An example of the forces influencing change in teacher education is the debate arising from the results of the TIMSS study. Germany scored in the middle of the field of all countries that participated in TIMSS II and III. The debate over possible reasons for this unexpected, unfavorable outcome include discussion of the obvious deficiencies of science teacher education programs, and as a result some research projects have been started to investigate the role of teacher education as a contributor to the poor achievement of students in science, and at making suggestions for improvement of science teacher education.

Historical Roots of Current German Teacher Education

Until the establishment of Germany as a country (1971) many states had individual school system structures and organizational solutions for the task of preparing teachers. The Prussian system was the most influential for both development of the school system and for establishing the system of teacher education for the individual states in the Federal Republic of Germany after World War II. The following discussion of the early developmental processes thus refer mainly to the Prussian system.

At the end of the eighteenth century the teaching profession was established as a dual school system consisting of the Gymnasium and the Volksschule. The Gymnasium system catered to a small group of students who were prepared for university studies, the precondition for leading roles in society e.g., political leaders, scientists and so on. The Volksschule (elementary school) system catered for the majority of students who were mainly from the lower class. Teacher education was structured quite differently depending on which of these two systems the teacher was being prepared for. For prospective Gymnasium teachers, preparation for teaching required university studies in that subject. In later years the requirement was increased to university studies in two subjects. There was no pedagogical education for transformation of teachers' scientific knowledge into teaching practice. A fundamental idea guided schools and universities at the beginning of the nineteenth century: Humboldt's conception of Bildung which was regarded as both the process of education and also as the goal of education. Bildung sought to help students to develop their mental and intellectual abilities so that they could become competent to decide and judge in a self-determined manner. This concept evolved from the European Enlightenment Movement, and required the development of the self-responsible, cosmopolitan person, contributing to his own destiny and capable of knowing, feeling and acting. Knowledge according to this concept had the function of supporting independent thinking and decision making.

Two aspects of this idea were influential in the design of teacher education:

- At the levels of the University and of the Gymnasium, knowledge was significant and justified on its own. Relationships of knowledge to applications of that knowledge were not essential to the goal or process of Bildung. Knowledge had to be acquired purely, for its own sake. Therefore, for prospective teachers, the knowledge to be acquired was the same as for prospective scientists, without reference to the differences in requirements for the different professions.
- The process of Bildung itself enabled participants to structure knowledge in such a way that it was teachable to others. Good knowledge and good Bildung were sufficient requirements for the ability to teach a subject at the Gymnasium.

The goal of university studies was not to educate teachers, but to educate citizens so that they succeeded in achieving Bildung. Some of those who decided to become teachers felt more like scientists than teachers, even when they worked in schools.

A second aspect of teacher education in the nineteenth century was in marked contrast to the conception of teacher education for the Gymnasium. In preparing to teach at the Volksschule, teachers were made familiar with the problems of practice in the schools at which they were assigned. The universities had no part in practice related studies. "For the Volksschule, the *volkstümliche Bildung* consisted of reading, writing, arithmetic, religion, a

subject called *Heimatkunde* dealing with local topography, culture and folk-
ways, very elementary forms of natural history, an introduction to the occu-
pational sphere on the level of craftsmen and workers, and in the rural areas,
farmers" (Terhart, 1998, p. 110). The education of teachers for the Volksschule, therefore, was structured quite differently from the education of
teachers for the Gymnasium. Prospective Volksschule teachers were not
required to study at a university. They were taught in special institutions just
the content that they had to teach later in classrooms. An essential part of
their studies was preparation for the demands of their profession. Pedagogical knowledge and teaching skills, therefore, had to be acquired to teach at
the Volksschule, where teachers not only had to convey subject knowledge
but also had to educate young children.

In principle this dualism within the school system as well as in the
organization and intentional structure of teacher education remained
unchanged until the first half of the twentieth century. Even the establishment of the Realschule (middle or technical school) did not put into question this dualism. Figure 7.1 shows the structure of the German school
system at the end of the twentieth century. The Gesamtschule (comprehensive school) is established only in some of the 16 states of Germany.

After World War II, several factors contributed to a development in
which old boundaries were crossed and different conceptions of education
came closer to each other. Among politicians, scientists, and pedagogues the
conviction grew that basic scientific knowledge is necessary for all people to

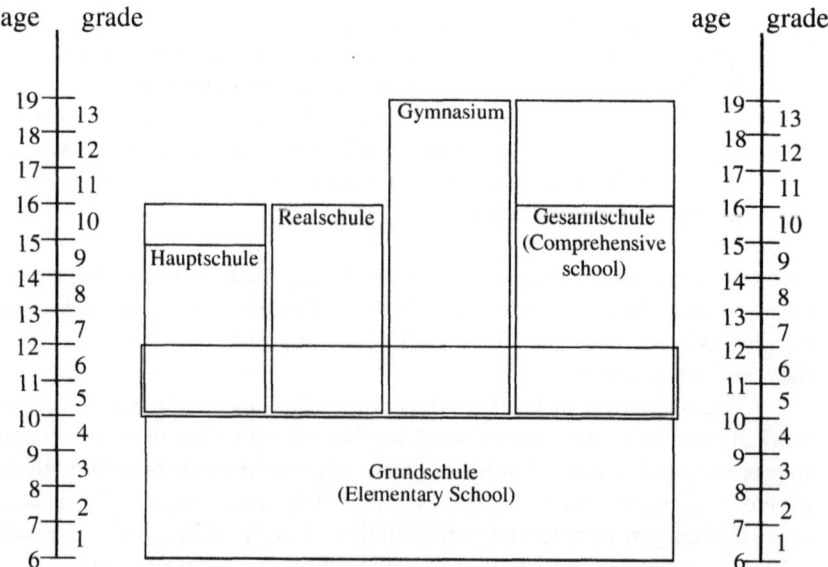

Figure 7.1. The school system in Germany.

withstand the requirements of the modern world. The stronger orientation toward scientific topics in the Volksschule resulted in teacher education programs that emphasized content-related studies more intensively than before. In addition to this, the goal of educating students toward self-determination and democratic attitudes required educators to strengthen this aspect of knowledge acquisition in their syllabi.

On the other hand, it became more and more obvious that teachers' expertise in a specific domain was not a sufficient precondition for successful teaching. Therefore, pedagogical elements were integrated into the prospective Gymnasium teachers' university studies. These developments—the integration of pedagogical studies for future Gymnasium teachers and the increase of content-related studies for the future teachers of the Volksschule/Realschule— were supported by the politically determined process of the establishment of the Gesamtschule (comprehensive school) in some German states in the 1960s. This school type removed the separation of different fundamental goals of schooling and promoted the differentiation between different school stages (lower and upper secondary level) instead of different school types. For teacher education, it generally became accepted that no fundamental differences exist between teachers of different school types or stages.

Science-Specific Aspects

The convergence of the general goals of the Gymnasium and the Volksschule/Realschule had a parallel history with the conception of science teaching. Science in the Volksschule was dominated by the idea of teaching topics that were useful to know. Knowledge was regarded to be useful when it contributed to the physical and religious health of students and allowed them to live a productive life. The convergence of goals in the last decades of the twentieth century led to a change in science teaching conceptions in which the natural sciences became more important in structuring the content of syllabi. For science teaching at the Gymnasium, conceptions of teaching moved toward criteria that stressed utility as at least equal to scientific principles in the syllabi.

Teacher education had to adjust to this development. At many universities the content and methods of general and subject-related educational studies are now less school type oriented and more school level oriented. The dominating idea now is that in the lower secondary level (grades 5–10, in some states grades 7–10) the problems of teaching and learning are quite similar but very different from that of the upper secondary level. As a consequence of this broadly accepted conception, curricular supplies, such as textbooks, are now supposed to be directed to students of the same level, regardless of the school type.

The same situation applies to research activities in science education. Investigations in mechanics, optics, or electricity do not differentiate between students of different school types. Learning problems, especially in the context of fundamental concepts such as voltage, force, energy, and so on, are similar for all students, regardless of their preknowledge or general ability.

Teacher education must prepare prospective teachers for this situation and must take into account the results of research in science education. Nevertheless, some universities offer special (low-level) subject courses (physics, chemistry, biology) to students who intend to teach in the Hauptschule/Realschule. In these cases Gymnasium students are integrated into the courses for diploma students. The question of whether or not special subject courses should be established for future teachers is discussed intensively. Students aiming at a professional teaching career answer "yes," whereas many scientists persist in the traditional teacher education conception in which prospective Gymnasium teachers have to learn physics (biology, chemistry) just as prospective scientists do.

The Organizational Structure of Teacher Education in Germany

Teacher education in Germany occurs in two phases. The first phase is located at the university, the second in schools. Figure 7.2 shows an overview of the fields to be studied and the time assigned to the two phases. On the right margin the average time is given which physics students need for their studies.

A severe problem for teaching as well as for research in science education is the organizational context in which science educators usually work. Normally they are affiliated with the department of their subject. Physics educators are members of the physics department, for example, and biology educators are members of the biology department. In the case of a vacancy the position of a professor of physics education, for example, is advertised by the physics department. A committee of physicists makes the appointment. Under these conditions it is not surprising that in many cases persons are appointed who have a general interest in teaching and learning but a strong physics orientation and are neither inclined nor competent to do research in science education.

University Studies: What Counts as Professional Knowledge?

It is difficult to describe in detail the structure and content of the 16 states' teacher education programs. Therefore, the following descriptions refer mainly to the program in Berlin, mentioning particularities when necessary. In most states two different careers for teachers exist, one for elementary and middle schools ("teacher") and one for the Gymnasium ("Studienrat"). Even those states that emphasize the integrating function of the comprehensive school and have teacher education programs oriented at school levels (and

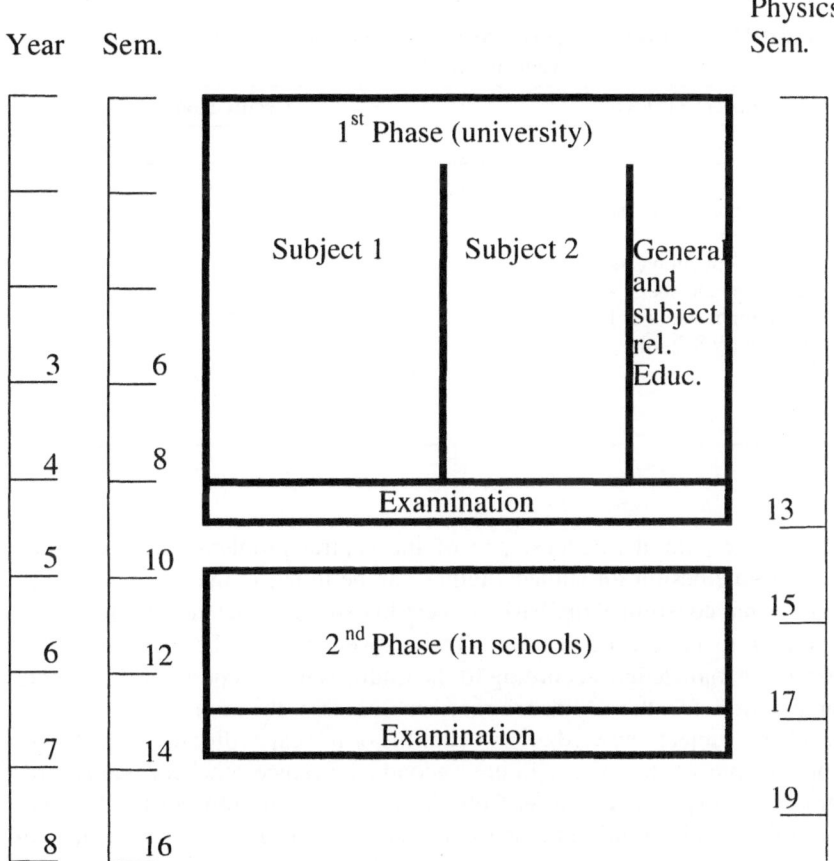

Figure 7.2. Organizational structure of teacher education in Germany.

not school types) have to maintain the traditional distinction between these different teaching careers because of the necessary mutual recognition of the certificates of the schools. Table 7.1 shows the credit-hours (semester-hours) required for "teacher" and "studienrat" in Berlin. Students normally take a content subject in their first semester followed by a combination of content and pedagogy in the second semester and throughout the rest of their program of study.

Subject Studies

The different numbers of credit-hours for the two teaching careers reflect the historical development in which the Gymnasium teachers take much more science than teachers for the other schools.

Table 7.1.

**Credit-Hours Taken by Students for Two Different Teaching Careers—
Teacher and Studienrat**

Subject Studied	Credit-Hours	
	Teacher	*Studienrat*
1st subject	54	72
Subject education (1st)	10	8
2nd subject	54	54
Subject education (2nd)	10	6
Education and social sciences	20	20
Educational sciences for Elementary School	12	
Total	160	160
	Teaching practice 4 weeks per subject	

In the natural sciences, one of the central problems concerning an appropriate design for subject studies can be found in the following question: What constitutes the basic subject knowledge a teacher has to master in order to be a good teacher? The answer stresses the significance of extensive knowledge, according to the traditional conception of the teacher preparation for the Gymnasium, meaning that the more extensive the teacher's subject knowledge the better his or her capability to teach. Therefore, in most universities future secondary science teachers' study programs do not decisively differ from those designed for physicists, chemists, or biologists. In addition to the traditional conception of the significance of a teacher's knowledge some scientists and politicians argue that in periods of unsure employment perspectives for teachers it is important for students to have the choice of becoming either scientists or teachers. Consequently, future teachers do not study their subjects from the perspective of their future profession.

Only a few universities offer special courses for prospective teachers. In these courses, students have to decide quite early which profession they plan to take up after their studies. One advantage of this model is that students can change between different teaching careers throughout their university studies. No reliable data exist about the time when students' decisions concerning their professional plans take place.

Table 7.1 also shows another aspect of the subject studies. Nearly all future teachers for secondary schools have to take studies in another subject, which can be chosen from a variety of school subjects. Physics students often choose mathematics, some of them chemistry. There are no connections between the two subjects; therefore, teaching students do not have the opportunity to gather experiences for interdisciplinary teaching approaches.

Science Education (Science Didactics)

Content and methods courses in physics education, biology education, and chemistry education are probably similar to the same programs in other countries. Two factors, however, characterize a difference from Great Britain and the United States. The first is the rich tradition of subject-related research on learning and teaching processes developed in the institutions of teacher training for teachers in the Volksschule/Realschule. The other factor is the science educators' affiliation to the department of their subject. This often results in a more content-specific orientation of the teacher education courses. Nevertheless, it is a broad consensus that teaching students should be able to analyze and reflect on goals, conditions, processes, and results of subject-related learning and teaching and to plan, design, carry out, and evaluate lessons.

Studies in physics (biology, chemistry) education often take place in three phases, which can be described according to their specific functions within the whole program of study:

1. An introduction which gives an overview and refers didactical topics to general results of pedagogical and psychological research. The main problems with students' learning science are discussed and illustrated with examples.
2. The basic skills necessary to prepare experiments and to give lessons are acquired. Principles of lesson planning are presented and discussed. This includes defining goals, decision making concerning methods, and applying criteria for using media.
3. In a third phase after the teaching practicum in-depth studies lead students to an elaboration of research-based knowledge.

In a "teaching practicum," which often takes place over a period of 4 weeks, students have the chance to teach on their own. During these activities students are permanently advised by mentors in school and supervised by university teachers (professors). The main goals of this period of practice halfway through students' university studies are:

- Student teachers become able to analyze the lessons they have observed or held themselves.
- Student teachers become qualified to understand the connections between different teaching decisions and to translate them into the planning process.

- Student teachers take part in procedures of assessment and get to know problems connected with them.
- Student teachers experience school life through their participation in conferences, school events, and class trips.
- Student teachers reflect on their decision to become a teacher.

Educational and Social Studies

This field of study has not developed a tradition that could be regarded as common to all universities. Therefore, not only the number of credit-hours but also the content offered to teaching students varies considerably between different university programs. All courses and lectures are offered by the department of education without any reference to subject-related topics. As a consequence of this disparate area of study, there is no integration of content and pedagogy during their programs of study and students have to do this on their own.

Examination

Teacher education in Germany is state controlled. The main regulations for the final examination are issued by the states' ministries of science and education. The content and focus of the final examination is determined by this body, which attempts to serve as a corrective agent to ensure a combination of content and pedagogy in the questions included in the final examination. The examinations are held by university members under the control of members of a state institution responsible for the organization of all examination affairs.

Students have to write a take-home essay. For Studienrat a scientific theme is given without any connection to school science. For other teachers a reference to school science is the norm. Presently this separation may be overridden in some states and is subject to change in others.

The Second Phase of Teacher Education

The second phase is a special characteristic of the German teacher education system. This phase is located in schools and can be described by the following factors:

- It takes place over a period of 2 years.
- Teacher candidates (Referendare) have to teach a reduced number of obligatory hours (about 5–8).
- They are already employed by the state and get about a third to half of teachers' initial salary.

Candidates have to attend three weekly seminars that are run by experienced teachers appointed by the school administration. In the main seminar general aspects of teaching and learning as well as legal and administrative problems are focused on. In two discipline-based seminars subject-related problems of teaching and learning are discussed in close contact with processes that happen daily in the experienced teachers' or candidates' classrooms. Topics worked on in the subject-based seminars are literature for teachers, experimental work, safety regulations, syllabi, planning schemes, goals of science education, teaching methods, media, assessment, and the like.

Experienced science teacher trainers who run the seminars and supervise the candidates have a great influence on the professional development of science teachers. With the determination of seminar topics and the setting up of standards of teaching they create a didactic frame that shapes candidates' pedagogical thinking. The importance of this phase for the actual and future practice of science teachers is significant; it is at this time that teachers should see the practical application of the theory they studied at the university.

One of the central problems from the perspective of the university is the selection of the experienced teachers as trainers (Fachseminarleiter, head of a subject-based seminar). There are no obvious criteria for this selection process. The fact that they are appointed by the administration indicates that they are acknowledged not because they are innovators but because they are respectable teachers who have shown to be able to stand the demands of teaching. It is not surprising that most of them are not inclined to look favorably on the first phase of teacher education, which they often regard as theory oriented and too far away from the conditions of practice.

The whole system of teacher education is thus divided into two parts: (1) academic studies focused on theoretical reflections with integrated practical periods and (2) practice-oriented studies with in many cases few references to academic foundations. Suggestions for reforming teacher education, therefore, aim at a strong cooperation between the two phases.

A second state examination finalizes the second phase of teacher education. Parts of this examination are:

- A written exposition on a teaching sequence done by the candidate, integrating a reflection on the teaching goals, the planning of the lessons, and an evaluative section.
- A lesson in each of the candidates' subjects, followed by a discussion in which the candidates are expected to demonstrate their evaluative conceptions.
- An oral examination in subject-related and general aspects of teaching and learning.

Teacher Inservice Education

All states in Germany have established institutions that offer courses, seminars, and the like to teachers who are willing to continue their professional development beyond the first two phases of teacher education. No official data regarding teacher engagement in these institutions exist, but in general the German inservice system is not regarded as a very successful one. One of the main problems concerns the legal framework of inservice education. Teachers are obliged to take part in further education by law, but there are no authorities controlly whether or not teachers fulfill their obligation. Because of their status as public servants any control or condition that could be imposed on them is not in accord with the principles of the public service. On the other hand, credits gained by teachers in further education do not improve their salary or their social career.

In reform periods in the last several decades inservice training courses were a part of school or curricular innovations (e.g., comprehensive school, interdisciplinary science). Presently the school system as well as the curricular situation is characterized by the idea to conserve the standards that have proved to be worthwhile. Therefore, the innovative aspect of inservice training does not play a significant role any more.

Perspectives of Teacher Education in Germany

Recently, a commission set up by the states and consisting of researchers, teacher educators, and administration officials presented a report on perspectives of teacher education in Germany (Terhart, 2000). This report summarizes the main developments that, in the view of the commission, should take place in the years to come:

- All institutions engaged in teacher education should regard this task as an integral one in which all three parts have their specific functions. The central recommendation of the commission is, therefore, to take into account the necessity of learning in the third phase of teacher education. This implies the change of institutional structures.
- The commission recommends keeping the organization at universities with studies in subjects, didactics, and educational sciences together with practical periods parallel to each other. But all parts should move toward a more intensive professional orientation.
- The second phase should undergo several modifications. A better coordination with the first phase is necessary. The selection of teacher trainers should be transparent and preceded by a qualification of those who apply for this position.
- Young teachers in their first years in schools should experience support and advice by colleagues and school administration.
- Competence and career should be linked in the professional development of teachers. For instance, salary should reward well-done work.

Not mentioned by the commission but a central problem is the very one-sided orientation of the study programs in science. The current situation in which prospective science teachers and prospective scientists in many universities have to study the same program, would result in a shortage of science teachers in schools.

TIMSS and Its Consequences

As mentioned in the introduction to this chapter, the results of the TIMSS study stimulated many discussions about the reasons for students' low scores. A multitude of factors have been held responsible. One of the numerous reactions to these results was the establishment of a research program financed by the central public funding organization in Germany (DFG, Deutsche Forschungsgemeinschaft): The Quality of School: Studying Students' Learning in Math and Science and Their Cross-Curricular Competencies Depending on In-School and Out-of-School Contexts. This program is guided by several theories. The current state of research shows that not only characteristics of the immediate learning environment are decisive for the quality of learning. Whether and how students perceive demands on learning and opportunities to learn depends also on factors related to the context. The climate or the profile of a school can be assigned to the school context. Parents and their relationship to school or peer groups of students are examples for the external context of school. The goal of the research program is, therefore, the following: "The systematic analysis of the impact of the different conditions to shed light on the complex interaction between school related and non-school related factors. Based on this analysis, BIQUA Bildungsqualitat von Schule (Quality of School Education) is to then suggest and evaluate theoretically well grounded interventions aimed at contributing to the increase of the quality of education in German schools."

TIMSS Video (Stigler et al. 1997) demonstrated that in different countries specific patterns exist that steer the design and processes of a lesson, including students' and the teacher's activities. These patterns, in the research program called "scripts," constitute the main factor of influence for the actual effects of science and mathematics teaching as well as for its further development. In light of this view of teaching, a systematic reconstruction of these scripts and investigations about their possible modifications and progresses are parts of the research program.

Teacher education plays an important role within this research conception because teachers can be regarded as leading factors in the bulk of context variables. Investigations with teachers in the third phase of teacher education promise to produce results that can be implemented in teacher education programs with a great number of teachers. Therefore a specific research project focuses on the question of how teacher education can contribute to an improvement of science teaching.

The project is in consonance with the results of TIMSS showing that not single characteristics of the teaching process or of a teacher's behavior are responsible for students' achievements, but the optimal accord of instructive variables in a consistent teaching conception and appropriate actions of the teacher well-suited to this conception. Figure 7.3 shows the essential ideas of the priority program. The factors to be stressed more intensively in future research projects are written in bold letters. In science, researchers have investigated many details of the processes of teaching and learning (media, methods, teaching skills) and their influence on students' learning, they have identified teachers beliefs about teaching and learning and their pedagogical content knowledge, and they have found many preconceptions held by students that impede their learning. The idea of the priority program BIQUA is that we should focus more on general patterns and scripts of the processes of teaching and learning and should take into account more the context of teaching and learning.

The current state of research suggests that we should regard teaching scripts as appropriate factors to describe the quality of teaching processes in science and mathematics. The teaching scripts designed and determined by teachers are formed by teachers' decisions, which are shaped by their pedagogical and psychological expertise. If students' achievements are to be improved and if changes of teaching scripts are possible measures for this improvement, then the identification and modification of teachers' competencies, which visibly influence teaching processes, are necessary preconditions for the success of such an endeavor.

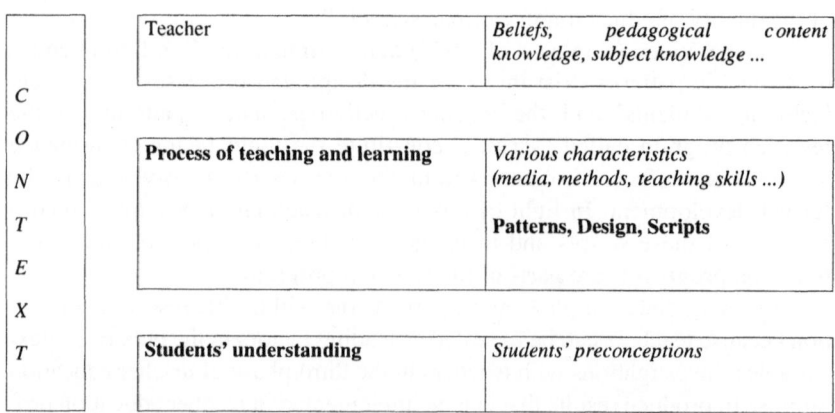

Figure 7.3. The schema for the German priority program BIQUA (Quality of School Education). Factors to be focused on in future research projects in teaching and learning in science (in bold).

The planned research project aims at investigating the conditions and effects of modifications of teachers' pedagogical and psychological expertise, which is acknowledged to be essential for the designing of lessons. The structure of this research project is as follows:

Initial analysis
- Identification of teaching scripts
- Identification of the teachers' expertise
    Conceptions about teaching and learning
    Conceptions about the nature of science
    Pedagogical content knowledge
- Identification of students' specific attitudes toward science education

Treatments to change teachers' expertise
- Reframing
- Reflecting-team
- Process consultation

Final analysis
- Analysis of the differential effects of the treatments
- Identification of teachers' expertise and their teaching scripts
- Questioning of the students

Three different treatments are planned to change teachers' conceptions about teaching and learning.

Reframing aims at restructuring the cognitive frame with which situations and processes are interpreted. For guiding conceptions of a reference frame new conceptions are offered that are suited to regard situations and resulting problems in such a way that teachers realize new possibilities to cope with the situation.

Reflecting-team: Reframing happens in group discussions in which a concrete teaching situation is regarded under the perspective of different interpretations of the teachers. The aim of the group discussion is to generate alternatives to the decisions made in the teaching situation. Because with this kind of reframing the status of teachers as experts plays a significant role, the alternatives generated in the group have an enormous degree of acceptance.

Process consultation focuses not on changing teachers' conceptions but on altering situation specific chunks of teachers' actions. Process consultation aims at interpreting these chunks and at offering new possibilities to act at central moments of a typical teaching process.

The research project is investigating the effects of these different treatments on teachers' teaching scripts and on students' achievements in science. In contrast to the United States and Great Britain, research in the field of teacher education is done only by very few groups in Germany.

The project described here focuses mainly on work done in the first phase of teacher education (Fischler 1994a, b; 1995; 1999).

DFG Priority Programm BIQUA (2000): The Quality of School: Studying Students' Learning in Math and Science and Their Cross-Curricular Competencies Depending on in-school and out-of-school contexts. www.ipn.uni-kiel.de/projekte/biqua/biqua-eng.htm

References

Fischler, H. (1994a). Concerning the difference between intention and action—Teachers' conceptions and actions in physics teaching. In I. Carlgren, G. Handal, S. Vaage (Eds.), *Teachers' minds and actions. Research on teachers' thinking and practice* (pp. 165–180). London: Falmer Press.

Fischler, H. (1994b). *Concerning the difference between intention and action.* Paper presented at the Annual Meeting of the National Association for Research in Science Teaching, Anaheim, CA.

Fischler, H. (1995). *Barriers on the way from theory to practice in physics teachers' education.* Paper presented at the European Conference for Research in Science Education, Leeds.

Fischler, H. (1999). The impact of teaching experiences on student-teachers and beginning teachers' conceptions of teaching and learning science. In J. Loughran (Ed.), *Researching teaching* (pp. 172–197). London: Falmer Press.

Klafki, W. (1998). Characteristics of critical-constructive didaktik. In B.B. Gundem & S. Hopmann (Eds.), *Didaktik and/or Curriculum—An International Dialogue* (pp. 187–200). New York: Lang.

Stigler, J.W., Hiebert, J. (1997). Understanding and improving classroom mathematics instruction. An overview of the TIMSS video study. *Phi Delta Kappan, 79*(1), 14–21.

Terhart, E. (Ed.). (2000). *Perspektiven der Lehrerbildung in* Deutschland. Weinheim and Basel: Beltz.

Terhart, E. (1998). Changing concepts of curriculum. From "Bildung" to "learning" to "experience"—Developments in (West) Germany from the 1960s to 1990. In B.B. Gundem & S. Hopmann (Eds.), *Didaktik and/or curriculum—An international dialogue* (pp. 289–300). New York: Lang.

Chapter 8
Science Teacher Education in Kenya and Cultural Extinction
How Can the Past Serve as a Link to the Future?

NORMAN THOMSON

These people must learn submission by bullets—it's the only school; after that you may begin more modern and humane methods of education.

SIR ARTHUR HARDINGE (IN LONSDALE, 1989)

Kenyans are engaged in a national debate regarding their educational system. While under British colonial rule a 7–4–2–3 system and curriculum were constructed and remained intact through Independence (obtained in 1963) until 1985. Since Independence, the national curricula and examinations have been continuously modified to enrich and align them with the Kenyan experience and environment (Republic of Kenya, 1964; TIQET, 1999). The government has remained cognizant that continued basic alignment with European standards is necessary for accreditation and transferability. However, interpretations by some stakeholders with regard to the overall relevancy of the formal educational curriculum, given Kenya's history and changing needs, has constantly posed problems (Bogonko, 1992; Republic of Kenya, 1976; Tsuma, 1998).

In 1985, the government restructured the education and training systems to make them more practical with the hope that students would emerge more self-reliant (TIQET, 1999). An 8-4-4 system was introduced (Eshiwani, 1993). The 8-4-4 system extended primary education by 1 year and placed an emphasis on learning literacy and numeracy. In addition, the curriculum intended to include a focus on development of students' technical and vocational skills. The revision was consistent with the Kenyan government's realization that the primary school leaving examination (CPE) was highly competitive and there remained a relatively low number of positions available for students to continue at the secondary school level. Postponing the terminal primary leaving examinations by 1 year would allow for an additional year of low-cost schooling.

But since the 1985 system reform, and despite education accounting for 40 percent of the national annual budget, the Kenyan government has lacked adequate financial resources and the infrastructure necessary to support large-scale systemic complementary teacher education preparation, material requirements, and inservice courses to successfully implement the development of science-oriented hands-on technical and vocational skills learning. Student performance in secondary-level examinations in science and mathematics has remained severely problematic. For example, in the 1999 national mathematics terminal examinations girls scored an average of 10 percent and boys 14 percent.

University faculties have been willing neither to justify lower academic entrance standards lost through dropping the highly selective 2-year A-level component of the British 7-4-2-3 system nor to add what they view as "remedial" courses. In addition, even though it has become more difficult for high school leavers to enter overseas universities, Kenya continues to suffer a "brain drain" of its most talented students. From a traditional African perspective, the younger students entering the country's universities have been viewed by elders as lacking maturity and self-discipline and are disruptive, that is, prone to protestations of hardships that lead to strikes and university dismissal.

Today, the 8-4-4 educational system is viewed by many Kenyans as a massive failure, and a new model with several optional strata, but fundamentally a return to a 7-4-2-3 system, has been proposed by a Commission of Inquiry, popularly known as the Koech Report (TIQET, 1999). Others view another major reform or reversion to the old 7-4-2-3 educational system as too costly. The current dilemma for Kenyans, then, is in the short term, "We cannot afford the cost of going back" and in the long term, "We cannot afford *not* to go back." However, as in many African countries, what has been lacking on both sides of the discussion is attention and regard to Kenya's little researched, appreciated, and potential contributions from its indigenous educational systems, which remain intact, extant, and locally important outside the sphere of formal education. Historically, the colonizers (religious and secular) and subsequent "development experts" have held most indigenous practices, including education, in little regard. And some Kenyans themselves must also undergo "decolonization of the mind" (Ngugi wa' Thiongo).

Kenya's dilemma is not unique. As Ki-Zerbo has stated, Africa continues to face a diametrical problem. First, "Africa is in serious trouble, not because its people have no foundations to stand on, but because ever since the colonial period, they have had their foundations removed from under them"; second, "there is generally a surprising lack of research to back up proposals for educational reforms" (cited in Gerdes, 1995, pp. 6–8). Seyni Kountché, former Niger president, has expressed similar sentiments: "Africa has been searching for a model of development. Extrapolations from the experience of other countries, or direct transplants, are often seized on because they seem to provide easy answers. In either case, an essential factor is neglected: the traditions and customs, so powerful in Africa, on which we have forged our civilization" (cited in Timberlake, 1986, p. 199).

A major issue now extending beyond the African continent and occurring on a global scale, but at a fundamental level of survival, is that nearly one-half of the world's 6,000 to 7,000 languages (2,400 are endangered and 50% of these are considered dying) are facing rapid extirpation (Brenzinger, 1992; Cox, 2000; Krauss, 1992; Wass, 1999), with the possibility that up to 90 percent will disappear in the twenty-first century (Weuthrich, 2000). These languages include indigenous knowledge systems in science. Africa has at least 1,000 distinct indigenous languages, and at least 200 of these are facing language death (Sommer, 1992). In their passing, several thousand years of science knowledge may become extinct. Since extinction is forever, a concern, responsibility, and role for science educators should be the preservation and promotion of indigenous science knowledge in Africa.

Throughout Kenya's history, Eurocentric domination has relegated indigenous science and science education to a point of being at risk of extinction. Colonialists established a myth that only through constant Westernization can high academic expectations and standards be legitimated and maintained. Current ideologies have done little to diminish the mystification of non-Western cultures. In this chapter, it is posited that science educators and teachers have as much responsibility in maintaining indigenous science knowledge and diversity in education, "edudiversity," as do biologists, indigenous cultures, and biodiversity (Cox, 2000; Wuethrich, 2000; Zerner, 2000). Given this setting for Kenya's current educational dilemmas, this chapter begins with a contextual cultural overview of Kenya and continues with a historical review of Kenya's educational system, including its administrative structure. Next follows a description of Kenya's current science teacher education programs, and the chapter ends with the question, "How can the past serve as a link to the future?" presenting perspectives of one ethnic group's traditional science education system.

A Cultural Overview of Kenya

Kenya is 582,650 square kilometers and lies astride the equator in East Africa. Kenya's national motto, *Harambee*, is Kiswahili for "Let us work together." Kenyan indigenous cultures include 46 major languages (population 29 million). English is the official language and language of instruction, whereas Kiswahili is the national language. Reflecting Kenya's cultural heritage, the national education motto is "Unity in Diversity." Kenya's coast is known for its Arabic cultural traditions and trade routes. Inland, one finds diversity in the bustling "time is linear and minute-oriented" Nairobi metropolis contrasting with the traditions of the "time is cyclical and infinite" nomadic peoples in the vast semi-arid northern areas. Literacy (15 years and above, reading and writing) is 78 percent (male, 86%; female, 70%) in a country where 42 percent of the people live below the poverty line (Central Intelligence Agency, 2000). The combination of geographical and historical cultural diversity in a relatively small country that intersects with today's global village presents a complex educational challenge for Kenyans. The educational system is centralized, but it must respond with education that is relevant and useful for all ways of life ranging from urban to nomadic.

An Overview of Education in Kenya since Colonization

No Native Tribes in the world have been subjected to such a forcing process as those of the Kenya colony. Nowhere has the whole system of civilization been dropped so suddenly and completely in the midst of savage races as here.

AN AMERICAN MISSIONARY QUOTED IN EDUCATION IN EAST AFRICA, 1924

In 1846, the German Church Missionary Society established a school at Rabai, near Mombasa, in today's Coast Province. This is considered the start of formal European colonial education in Kenya. The school's primary purpose was to promote evangelism, but as education developed, it became an instrument to produce skilled labor for settlers' farms and clerical staff for the colonial administration. Education during the colonial period was racially stratified. There were separate schools and curricula (and much superior resources) for the Europeans (Eshiwani, 1993; Sifuna, 1990). Asian and Arab systems were next in status. The Colonial government left the building, managing, and supervision of the few schools for Africans in complete control of missionaries. Winston Churchill, (1962) after an official visit to Kenya in 1907, stated: "[D]iscipline, careful education, sympathetic comprehension are all that are needed to bring a very large proportion of the tribes of East Africa to a far higher social level than that at which they now stand." British administrative policy remained with "indirect rule" (in contrast to the "direct rule" utilized by the French, Belgians, and Portuguese). "Indirect rule" is one of adaptation; that is, the British took advantage of the existing structures of African societies for administration of policy, including education. Its disadvantage was that it maintained a status quo based on stereotypical perceived competencies of culture, class, and race. The colonized could never become British despite the fact that some of the colonized African societies were far more democratic than the monarchy. The British left the role of evangelism and education ("civilizing") to missionaries. In turn, the missionaries viewed their role as one of teaching basic reading skills, the catechism, and the development of vocational skills. Vocational skills were deemed important by the colonial administration because the outcome would be a wage-based labor-dependent African society. Labor served as a means for extracting natural resources and producing goods and allowed the British to maintain political, economic, and social control, providing justification for their presence. In irony, but in a different time and context, Kenya's 8-4-4 system is a return to an emphasis on skills development.

In 1920, Kenya changed from Protectorate to Colony status and, in 1926, the first academic secondary school (Forms I–IV) was established by missionaries for African males, Alliance High School. The chief subjects of the first 2 years were English, arithmetic, science, agriculture, and art, whereas the final 2 years were for vocational training in the field of teaching, medicine, agriculture, and forestry (Stabler, 1969). Later, missionaries also established training colleges for teachers and provided funds for most recurrent costs, with only a small subsidy from the colonial government in the form of grants. However, missionaries were untrained education professionals who promoted interdenominational rivalries between ethnic groups and focused on religious conversion. Some Africans perceived the missionaries as an integral part of the European presence—agents of colonization and oppression. They saw the contradictions in white missionaries preaching love and sharing, while their religious parishioners, the white settlers, were imposing hut taxes, confiscating traditional land and sacred forests, and holding uncompromising interpretations with regard to differing metaphysical beliefs.

Until Independence in 1963, there were, therefore, great disparities in educational opportunities perpetuated not only between races (see Willinsky, 1998), but also between the different regions and ethnic groups. Throughout the colonial period, stress was placed on labor-intensive technical and vocational education for Africans; from the Fraser Report of 1909, which first recommended an industrial curriculum as the basis of African education, through the 1949 Review Commission under Archdeacon Leonard Beecher, the objective was consistently to enhance the Africans' "suitability as laborers and craftsmen on the settlers' farms."

At Independence (1963) the colonial educational legacy posed many problems of quantity, quality, and relevance for the new government. Since then, Kenya has made enormous investments in education. The new government had to reorient educational policy not only to make it more relevant to the needs of a new nation, but also to expand access to education. Initially, the educational system remained intact, but Kenyans were needed for roles of leadership in government, academics, and the business sectors. The educational system underwent minor restructuring as the curricula and assessments were Africanized. Substantial expenditures for education have occurred, from a 10 percent gross national product investment in 1964–1965 to about 38 percent in 1990–1991. But the most profound change has been the 1985 replacement of the British 7-4-2-3 system with a Canadian/American 8-4-4 system. The change was intended to address increasing demands for an economy that was requiring technically and professionally qualified personnel. But with more than 6 million young Kenyans enrolled in various educational institutions, the 8-4-4 system is now seen as an unanticipated national disaster. The cost of conversion from a discipline-oriented education to a technology-based system was neither fully anticipated nor appreciated. And the anticipated outcomes of the new system have fallen short of the once acclaimed goals.

Kenya's Administrative Educational Structure

Kenya's formal educational system has always been centralized and administered through a Ministry of Education. Several institutional structures conduct and administer the system at national, provincial, district, and local levels. Nationally, the Kenya Institute of Education (KIE) consists of full-time education professionals who lead grade level and subject matter panels (representative school teachers, university professors, businessmen, and parents) that develop course syllabi, ensure that textbooks are written and published by the Kenya Literature Bureau, and develop continuing education programs; an Inspectorate that administers professional development workshops and monitors the quality of teaching/instruction through school and in-class inspections; Kenya National Examinations Council (KNEC), an autonomous body that develops and evaluates school and student performance on terminal and qualifying national examinations; the Teachers Service Commission, whose charge is to interview, hire, and assign teachers to schools throughout the country to ensure equity, teachers; from a cadre of primary, secondary, university and college, and technical schools, whose role is to conduct the education of students. Schools vary with respect to public/partial government support/private (religious and secular), boarding/day, and male/female/mixed. Prior to receiving approval by the Kenyan Parliament, Kenya's national syllabi and examinations are reviewed externally (by the Ministry of Education and Examination Boards) to maintain international standards.

Preprimary and Primary Education

Before 1980, preprimary education for children between 1 and 6 years of age was exclusively the responsibility of local communities and nongovernmental organizations. The government assumed responsibility for preschool education in 1980 and now includes training of preschool teachers, the preparation and development of the curriculum, and preparation of science teaching materials. Local communities and nongovernmental agencies meet the development of preschool units and the cost of teachers' services. The number of children attending preprimary units in 1990 was about 800,000, while the number of preschool teachers was about 20,000.

Primary education is considered the first phase of the formal education system. It usually starts at 6 years of age and lasts 8 years. The main purpose of primary education is to prepare children to participate fully in the social, political, and economic well-being of the nation. The primary school curriculum has been designed to provide a more functional and practical education to cater for the needs of children who finish their education at the primary school level rather than for those who will gain access to a secondary education.

Prior to independence, primary education was almost exclusively the responsibility of the community's concerned or nongovernmental agencies such as local church groups. Almost all primary schools in the country are now in the public sector and depend on the government for their operational expenses. The government provides teachers and their salaries and finances a milk scheme for all students and a feeding program for students in the semiarid parts of the country. Government expenditure on school supplies and equipment (except science) are minimal, as fees levied on parents by local parent teacher associations finance these. Parents are responsible for construction and maintenance of schools and staff housing (many schools provide housing at nominal rent to the teacher). Since Independence, almost all primary schools built and equipped have been the result of *harambee*, or local self-help efforts.

There has been a remarkable expansion in primary education, both in terms of the number of schools established and in the number of children enrolled, over the past three decades. At independence, there were 6,056 primary schools with a total enrollment of 891,600 children. At the same time, trained teachers numbered 92,000. In 1990 there were over 14,690 primary schools, with an enrollment of slightly over 5 million children and with nearly 200,000 trained teachers. In addition to the expansion in the number of primary students enrolled, there has been a significant improvement in the participation of girls in education. At Independence, only about one-third of enrollment in primary schools was girls; today it is nearly 50 percent.

But the lack of adequate high school places leads to cutthroat competition in the Kenya Certificate of Primary Education (KCPE) national examinations. In 1999, only 198,260 (44%) of the 446,539 KCPE candidates were admitted to Form One at the secondary level. The coveted national schools admitted only 1,600 (0.008%) candidates, while the second-tier provincial schools took 39,000 (20%) students. The irony of the situation is that one criticism of the former CPE is that it caused undue stress on the pupils by testing only three subjects. KCPE assesses 14 subjects in seven papers, unlike CPE, which had only three papers. The national mean score is about 45 percent of the 700 possible marks. Continuous assessment is not used as a measure of performance, and the current syllabi and examinations are thought to encourage memorization.

Secondary Education

Secondary school education usually starts at 14 years of age and, following the introduction of the 8-4-4 system, lasts 4 years. The current secondary education program is intended to focus on meeting the needs of students who terminate their education after secondary school. In this context, the new secondary school curriculum lays greater emphasis on job-oriented courses, such as business and technical education. However, the government has been unable to meet the financial requirements for the transformation of schools from an academic orientation to a hands-on practical or technical education. Transformation requires a locally relevant curriculum, qualified teachers, and instructional space and materials. Some schools prefer a focus on agriculture, whereas others are interested in large and small business industries. Any change represents a major investment and commitment from the stakeholders. There remain two types of secondary schools in Kenya, public and private. Public secondary schools are funded by the government or communities and are managed through a board of governors and parent teacher association. Private schools, on the other hand, are established and managed by private individuals or organizations, especially religious organizations.

There has been an increase both in the number of secondary schools and in student enrollment in response to the rapidly increasing number of primary school leavers seeking entry to the secondary level. In 1963, there were only 151 secondary schools with a total enrollment of 30,120 students. Today, there are nearly 3,000 secondary schools with an enrollment of 620,000 students. Of this total, slightly over 40 percent are girls. Class size typically ranges between 40 and 50 students.

The rapid expansion at the secondary level has mainly been the result of the *harambee* movement, which has led to the establishment of numerous community secondary schools. Secondary students do 10 subjects for 4 years and must sit examinations during a 6-week summary period at the completion of Form Four. Most subjects have two examination papers, a $2\frac{1}{2}$-hour theory paper and a $2\frac{1}{2}$-hour practical paper (a total of 50 hours of examinations). However, as with leavers of primary school, the number of secondary students qualified to enter the next educational level is small, about 20,000 (12%) per year. In addition, continuous assessment is not included as a final measure of performance and it is thought that the current syllabi and examinations encourage memorization. The school year is divided into three terms of 13–14 weeks. The last 2 weeks of each term throughout a student's secondary education are devoted to doing comprehensive "mock" examinations, which emphasize the term's current topics and also sample all previous syllabus material. Thus, in most instances, students spend 4 years continuously preparing for the high-stakes terminal examinations.

University Education

The first step toward the introduction and development of university education in Kenya was taken in 1961 when the Royal College in Nairobi was elevated to university college status. The college entered into a special arrangement with the University of London, which enabled it to prepare students for the university's degrees. With the establishment of the University of East Africa in 1963, the Royal College became the University College, Nairobi. The other constituent colleges of the University of East Africa were Makerere in Uganda and Dar-es-Salaam in Tanzania. The University of East Africa continued operating until 1970 when the University College of Nairobi attained independent university status.

Apart from the establishment of Kenyatta College (with a focus on teacher education) as a constituent college of the University of Nairobi in 1970, the latter remained the only university in the country until the mid-1980s. In the preparation and induction of science teachers an emphasis was placed on knowledge in a major and minor science content area with an overall emphasis on mathematics. Since the mid-1980s, there has been a tremendous expansion in universities in response to the high demand for university education in Kenya. The country now has four public universities, with the most recently established universities giving greater emphasis to technology and science-oriented degree programs. In addition to the four public universities there are 10 private universities in the country offering a range of degree programs. They are supervised and controlled by the Commission for Higher Education.

Science Teacher Preparation in Kenya

Primary School Teachers

Kenya has several teacher training colleges that prepare primary school teachers in a 2-year, six term program. The course of study includes all 13 primary school subjects, of which science is one. To obtain a primary teacher's certificate a candidate must pass in at least eight subjects (continuous assessments and final examinations) and must pass practical teaching. Thus, a candidate may or may not pass the science component for teaching qualification. Time spent on preparation to teach elementary science topics is 236 hours (Table 8.1).

Table 8.1.

Science Topics Covered and Allocation of Hours Spent in the Preparation of Teachers at the Primary School Level (P1–P8).

Unit	Topic	Time Allocation (h)
1	What Is Science?	20
2	Weather	18
3	Teaching/Learning Materials	8
4	Primary Science Syllabus	28
5 & 13	Demonstration Teaching	20
6	Environment	16
7	Living Things	34
8	Health Education	6
9	Properties of Matter	22
10	Energy	28
11	Simple Machines	18
12	Evaluation	18
Total		236

Secondary School Science Teachers

Kenya Science Teachers College (KSTC) was established in cooperation with the government of Sweden in 1965 to prepare nongraduate secondary school science teachers. Initially, it accepted qualified O-level leavers and upon successful completion of a 3-year program future teachers were awarded an S1 certificate that allowed them to teach two science subjects at the secondary level. In 1980, KSTC began a 2-year Diploma in Science Education for A-level science leavers and in 1992 reverted to a 3-year program offering a Diploma in Education and Diploma in Science and Technical Education in response to the 8-4-4 system (Table 8.2). Students admitted to the program may be selected for six subject combinations from two areas to include mathematics, physics, biology, or chemistry. In addition, all students are required to take subjects in education, physical education, communication skills in English, environmental science, a workshop course in preparing laboratory materials, library science, and guidance and counseling. All students leave qualified to teach at least two science subjects and physical education and are to be self-reliant in designing and building basic science laboratory equipment. KSTC has 12- to 13-week terms, and each student has a total of 2,376 contact hours (Table 8.3).

Moi University

In 1963, at the time of Kenya's Independence, there were only 500 undergraduate students in the country's single university in Nairobi. By 1981 the number had increased to over 9,000, with an additional 7,000 pursuing degree

Table 8.2.

The Sequence of Courses All KSTC Students Follow

Year 1

Term 1	Term 2	Term 3
2 Science Subjects	2 Science Subjects	2 Science Subjects
Education	Education	Education
Physical Education	Physical Education	Physical Education
Communication Skills in English	Communication Skills in English	Communication Skills in English
Environmental Science	Environmental Science	Environmental Science
Workshop Course	Workshop Course	Workshop Course
Library Science	Library Science	Library Science
Guidance and Counseling	Guidance and Counseling	Guidance and Counseling

Year 2

Term 4	Term 5	Term 6
2 Science Subjects	2 Science Subjects	2 Science Subjects
Education	Education	Education
Physical Education	Physical Education	Physical Education
Communication Skills in English	Communication Skills in English	Communication Skills in English

Year 3

Term 7	Term 8 & 9
2 Science Subjects	Teaching Practice
Education	2 Science Subjects and
Physical Education	Physical Education

Table 8.3.

Required Time Allocation of Contact Hours for KSTC Students in Science Teacher Preparation

Subject	Contact Hours (h)	Percentage of Time
Biology/Chemistry/Mathematics/Metal Work/Physical/Woodwork	732	31
	732	31
Education	324	14
Physical Education	216	9
Communication Skills in English	144	6
Environmental Science	72	3
Workshop Course	72	3
Library Science	48	2
Guidance and Counseling	39	2
Teaching Practice	480	20

courses in overseas university because of a lack of places in country. In 1984, Moi University was established in a rural setting 400 kilometers from Nairobi; its mission is to focus on teaching, research, and service in science and technology with an orientation toward social, cultural, and agricultural aspects of Kenyan life. The university offers a 4-year Bachelor of Education Science degree program in preparation for teaching the secondary-level sciences. Students are required to have a combination pair of a major (chemistry, physics, mathematics, or zoology/botany) and a minor (physics, biology, mathematics, geography, or chemistry) subject area. During the first year of study students are tracked with other science majors in either a physical (math-physics-chemistry/MPC) or biological (maths-bot/zoo [biology]-chemistry/MBC) course combination in the Faculty of Science (Table 8.4). In addition, foundation courses with an emphasis on Kenya's heritage and experience are taken through the Faculty of Education and the Institute for Human Resource and Development. Students are expected to learn contemporary educational philosophies, practices, and technologies, but also to have a clear understanding and appreciation for Africa, especially Kenya's contextual political and social aspirations.

Following the first year of studies, students specialize in their major and minor content areas (Faculty of Science), but have in common their professional education courses with other disciplines (Faculty of Education and the Center for Human Resource Development). At completion of their studies, students will have completed 78 credits in the major subject area, 21 credits in the minor subject area, 63 credits in professional education courses, and 21 credits in human resources and development, with a total of 188–194 credits

Table 8.4.

Courses Required During the First Year of University for Prospective Secondary School Science Teachers

Subject	Semester I (Credits)	Semester II (Credits)		
	MPC	MBC	MPC	MBC
Botany	—	3	—	3
Zoology	—	3	—	3
Chemistry	3	3	4	4
Physics	4	—	4	—
Mathematics	3	3	3	3
Education	3	3	3	3
Human Resource Development	6	6	6	6
Computer Science	—	—	3	3
Total Credits	19	21	23	25

Two track combinations are offered: Mathematics, Physics, and Chemistry (MPC) or Mathematics, Biology, and Chemistry (MBC).

Table 8.5.

Curriculum and Semester Credit Hours Required for Prospective Secondary School Science Teachers during Years 2–4

	Semester I (Credits)	Semester II (Credits)	Term 2 (Schools)
Year 2			
Major Subject	9–12	9–12	4 weeks of practice
Minor Subject	3–6	3–6	teaching in a secondary
Education	9	9	school
Human Resource Development	3	0	
Total Credits	27	24	Max = 51
Year 3			
Major Subject	12	12	12 weeks of practice
Minor Subject	3	0	teaching in a secondary
Education	9	6	school
Human Resource Development	0	3	
Total Credits	24	24	Max = 48
Year 4			
Major Subject	12	12	
Minor Subject	3	3	
Education	9	6	
Human Resource Development	0	3	
Electives (2)	0	1 X 2	
Total Credits	24	26	Max = 49
Degree program total maximum		188–194	

for graduation (Table 8.5). All education students are required to have 4 weeks of teaching practice at the end of the second year and 12 weeks at the end of the third year of study (thus, teaching practice takes place during the second term of the secondary schools). Teaching practice is pass/fail, but it is neither taken as a course for credit nor is it an officially examined course and, consequently, does not count toward graduation credits. Moi University has 15-week semesters; each science education student is required to have a total of 3,130 contact hours (Table 8.6).

Kenya—A History of Education

Just as Kenya serves as the cradle of human origins, so too does it also most likely serve as the cradle of education, including science education. However, traditional educational systems were abhorred by the colonialists and,

Table 8.6.

Required Time Allocation of Contact Hours for Moi University Students in Science Teacher Preparation Program

Subject	Contact Hours	Percentage of Time
Major Subject	1170	37
Minor Subject	315	10
Education	945	30
Human Resource Development	315	10
Teaching Practice	384	13

until recently, have been neglected by Kenyans (Njoroge & Bennaars, 1994). Kenya is characterized by Mazuri (1986) as sharing a triple heritage of worldviews: its indigenous heritage and two major exogenous influences, Islam and Westernization. In the seventeenth century, as Western natural philosophy was establishing a role for itself as a legitimate way of understanding the natural world, Francis Bacon wrote that the most noble ambition for man was "to establish and extend the power and dominion of the human race over the universe. . . . We cannot command nature except by obeying her and understanding her" (cited in Aikenhead, 1994). During the eighteenth century, in the quest for access, exploitation, and control of natural resources in Africa, western European nations felt it obligatory that global "power and dominion" include indigenous peoples (Mazrui, 1986). Consequently, traditional African beliefs, values, and identities were often relegated to a status of existing as primitive antiquities needing replacement. Replacement has had two natural allies in the guise of missionaries: Traditional metaphysical beliefs were challenged through religious education (Mbiti, 1969; p'Bitek, 1984), and epistemological beliefs were addressed via secular education (Bogonko, 1992; Njoroge & Bennaars, 1994; Shiundu & Omulando, 1992).

A Brief Description of One Indigenous Kenyan Educational System

The following is a work in progress documenting the traditional educational system of the Keiyo people (Thomson, 2000a). The Keiyo (1990 population about 120,000) are one among eight Kalenjin ethnic groups in Kenya. Specifically, the Keiyo people live in Keiyo District of Rift Valley Province in northwest Kenya and occupy an area about 100 kilometers in length and 20 kilometers in width. The district is traversed by the Kerio rift escarpment rising from 2,500 to 5,700 meters above sea level.

The first documented encounter of Keiyo with a European explorer was the passage of Joseph Thomson (1887) in November 1883. On his journey in search of the great lake, to be named "Victoria Nyanza," he met the Elgeyo in Kerio Valley, traversed the Kerio escarpment, and continued westward on the

Uasin Gishu plateau. In 1891, the German Emin Pasha Expedition tried to follow the same route and encountered Keyu resistance on the escarpment (Peters, 1891). The killing of Keyu warriors initiated prolonged mistrust of Europeans.

Keiyo culture has an educational system based on lifelong learning in a society structure of cyclical "age groups." All Keiyo people have equal social status; however, each age group has specific responsibilities that synergistically serve the welfare of the community. In Keiyo education, each age group is not only expected to conduct their responsibilities, but is also being prepared for future age groups with an emphasis on the next immediate age. Keiyo education emphasizes consensus building through discourse. Traditional Keiyo education places value in *ng'omnon* (knowledge) through *ipwaat* (to think, remember, recall) and *inai* (coming to know or understand).

In Keiyo culture, learning occurs in informal contexts, through work (*keboisie*) or play (*keureren*), and in formal (*tum*) contexts. In addition, the Keiyo make a distinction between the teachers involved. Within the processes of informal education, at times a teacher (*kanetindet*) may spontaneously engage learners as a situation arises, but other learning times are centered on particular tasks, time of day, season, or related to ceremonies. Keiyo education requires that a comprehensive curriculum be learned, including language, beliefs, proverbs (*kalewenut*), games (*ureriet*), storytelling, science, agriculture, mathematics, and teaching children their societal and individual tasks and responsibilities.

Formal knowledge transfer to children (*tumdab lagok*) is most emphasized as a part of initiation from adolescence into adulthood, a period that traditionally lasted several years (*tumdab torusiek*, learners' [adolescents'] bridge to adulthood). Highly regarded individuals are chosen as these teachers (*motiren*), and a distinction is made between male (*kanetindetab murenik*) and female (*kanetindetab chepyosok*) teachers. Learning in either informal or formal contexts can occur through watching (*okertoi*) or demonstration (*kerub*), participatory modeling or apprenticeship (*fuata nyayo*), investigation (*kekit*), or teaching one's self (*inetgei*). Assessment of learning and understanding is highly contextual. Tests (*tiem*) may result in success (*sulda*) because of either effort or luck (*bor*), and mistakes (*leel*) may lead to failure (*ache keiput*).

Keiyo Teaching of Chance, Probability, and Problem Solving

The mathematics of chance, the mathematical description of randomness, is called probability theory. Probability describes the predictable long-run patterns of random outcomes and can be expressed as a fraction, decimal, or whole number ratio ranging from "impossible" to "certain." The probability of an outcome is the proportion of trials in which the outcome occurs over the total possible range of occurrences or a very long run of trials (Moore,

1994). Prediction and hypothesis testing about the probability or occurrence of an event(s) form the basis of hypothesis construction and decision making relevant to possible outcomes.

Discovery and problem solving in science can be viewed as having stages and strategies, utilized in hypothesis construction, use, and revision (see Thomson, 1993). A hypothesis is a proposition or statement used to create, examine, and explain probabilistic patterns for relationships among independent and dependent variables. Keiyo language includes concepts of probability: *aaoksei* (likely), *maaoksei* (not likely); contextual relativity as to space and time: *nyi, anan nyi* (this, or this) versus *nyi, anan nyin* (this, or that); and contextual prefixes that begin propositional/conditional statements: *Ngi . . . , i . . .* (if, then). (Note: Keiyo is an oral language without a translated dictionary or grammar.)

In secondary school, the teaching of transmission genetics to students has typically been a starting point for learning genetics and serves as an important foundation for learning biological concepts of diversity, inheritance, and evolution (Dobzhansky, 1973; National Academy of Sciences, 1998). Consequently, transmission genetics has been included in the syllabi for almost all secondary schools (National Research Council, 1996; Kenya Institute of Education, 1997). Transmission genetics, a domain based upon probabilistic events, is an important but one of the more difficult topics to learn in high school biology (Finley, Stewart, & Yarroch, 1982). Thomson and Stewart (1985) proposed an algorithm for solving typical high school biology textbook problems to help make genetics instruction more explicit and meaningful to students. Since that time an open-ended computer simulation of classic Mendelian genetics problems, Genetics Construction Kit (GCK) (GCK, BioQuest, 1999), offers a probabilistic problem-solving environment. Using GCK, Thomson (1993) identified experts' problem solving stages and levels of strategies based on Darden's (1991) framework for theory change in science. From a different perspective, Glynn et al. (1995) have used analogies as a way to teach students about transmission genetics. Thus, given the sociocultural and constructivist views on learning, can analogies be used in the realm of understanding probability and problem-solving strategies that may be shared by a traditional game and transmission genetics?

Traditional Cattle Raiding

Complex strategies were used for the preparation and conduction of successful cattle raids. Planning was initiated through clan/group consensus, and cause (independent variables) could either be a shortage of food (long-term drought) or a need to increase the number of cattle following disease (e.g., Rift Valley Fever; see Linthicum et al., 1999). Groups of six to eight morans were formed based on variables such as speed, dependability, experience, past success, and the ability to create comprehensive raiding strategies.

Numerous groups would go in various directions on independent reconnaissance missions. Variables considered for effectively scouting and raiding a particular site included number and quality of cattle, days of travel involved, strengths in offense/weaknesses in defense, distances from water and home, time of year/month/day, and overall site vulnerability. Following reconnaissance, the clan would meet and discuss possible events, options, and outcomes prior to making a decision for a raid. Following a successful raid, cattle would be equitably distributed among clan members, but individual cattle could be selected for particular genetic traits. Furthermore, each clan member would selectively distribute his cattle to other *beutab tuga* (homestead with cattle) for breeding and so that raiding or disease at any one *beutab tuga* would not result in the loss of one's total herd.

Kechui, *a Game Based on Traditional Cattle Raiding*

Kechui is a simulation game played on a four-row, eight-column matrix usually by making 32 holes in the ground (Figure 8.2). The game is popular and spontaneously played by children outside classrooms. Each player's side consists of 16 (4 × 4) holes (*beutab tuga*) and represents an ethnic group of cattle raisers. Each player places an agreed-upon but equal number (usually 4) of Sodom Apple/*Solanum incanum* fruits (cattle) in each *beutab tuga*. Consequently, each player begins with as many as 128 cattle. In the outer row, in one corner, and diagonally at opposite ends of the matrix a home *beutab tuga* is chosen. This *beutab tuga* is an area safe from any raiding. Each time a player passes over this site a bull/cow becomes safe. During a play, cattle are collected at one site and the dropped one per hole at a time. Direction may be changed and rules determine when you may raid and when your play is completed. The game's goal is not only to get all your ethnic group's cattle into your home *beutab tuga*, but also to successfully raid the neighboring ethnic group's cattle. Each play takes into account collecting your own ethnic group's cattle, raiding the opponent's cattle, and passing over your home *beutab tuga* as often as possible. For each play, strategies and probability are used to develop an optimum series of moves. During a game more experienced (winners) players changed direction more often than less experienced players (losers) and utilized several strategies based on an understanding of probability.

Problem Solving in Transmission Genetics

Kenya's national mathematics (Form Three) and biology (Form Four) syllabi recommend coin and die tossing and counting seeds in pods to introduce probability (Kenya Institute of Education (KIE), 1992) with no reference to contextual cultural history. The scores of males ($M = 10.5$) and females ($M = 6.5$) were not statistically significant different (Mann-Whitney Test, $p > .05$) in solving past examination questions. In solving genetics construction

home *beutab tuga*

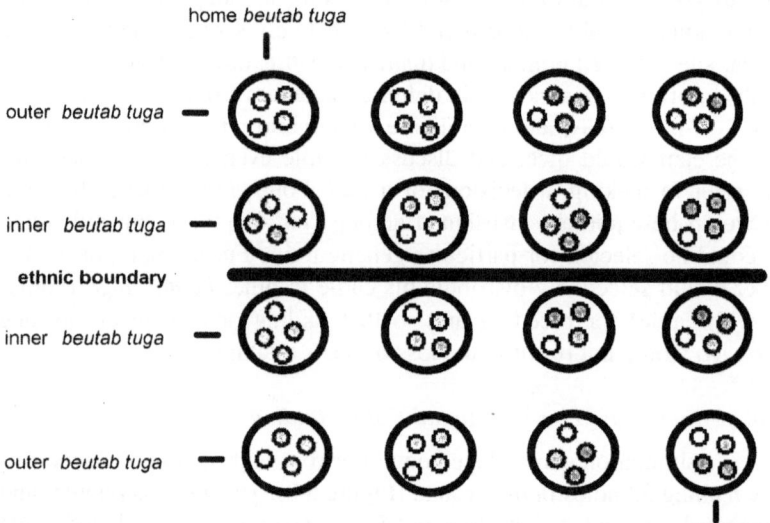

Figure 8.1. The game of Kechui is played on a 4 × 4 matrix of holes. A player's side consists of 8 beutab tuga representing one ethnic group of cattle raisers. Each player or side places an agreed upon and equal number (usually 4) of Sodom Apples (cattle) in each boma. A home beutab tuga is usually chosen diagonal to the opponents. The goal is to not only get all your ethnic group's cattle into your home boma, but to raid the neighboring ethnic group's cattle as well.

kit (GCK) problems three students solved both problems, 12 solved one or the other (complete dominance = 6, codominance = 6), and one student solved neither. Students approached problems as textbook-type problems. Most students succeeded only if either a 3 : 1 (simple dominance) or 1 : 2 : 1 (codominance) ratio was obtained through a chance cross. In general, the students expressed no understanding of probability except when ratios were stereotypical textbook numbers. For example, students assumed and concluded that any time a particular phenotype had a greater number of individuals the causal gene was dominant. In addition, rather than using a tree diagram or matrix (Punnett square) as advocated by geneticists (Russell, 1996) current Kenyan secondary textbooks (KIE, 1997; Soper & Smith, 1995), teachers, and students use lines to connect gametes to offspring. Using lines, students were unable to use this technique to follow more than one generation of organisms. GCK simulations and real problem-solving environments require generational knowledge construction. Interview data reveal that successful problem solving was more a consequence of chance outcomes than a probabilistic understanding of genetics.

Keiyo students used and demonstrated an understanding for concepts of probability in Keiyo while playing the matrix game of *Kechui*. However, in solving textbook and computer-based transmission genetics problems in English, little understanding of probability was demonstrated. Current Kenyan secondary mathematics and science textbooks make no reference to indigenous cultural knowledge or related activities. In addition, Kenyan textbook and teacher-based genetics instruction do not include matrix representations. Language may be one factor limiting knowledge transfer. Choice of analogical models used to introduce concepts of probability may be another.

Summary and Implications

As was mentioned before, a major issue now extending beyond the African continent and occurring on a global scale is the possibility that up to 90 percent of the world's languages will disappear in the twenty-first century. Since these languages include indigenous knowledge systems in science, a concern, responsibility, and role for science educators should be the preservation and promotion of indigenous science knowledge in Africa. In the twenty-first century, increased Eurocentric domination of cultures through processes of globalization places a plethora of indigenous science education systems at risk unless a concerted effort is made to document and legitimize not only indigenous ways of knowing but also what is known. Kenya, with over 40 major cultural groups, provides just one example of a "nation at cultural risk." Current Kenyan education has recognized international structure and standards, but 10,000 years of science education is being forgotten and needs to be incorporated into the syllabi, teaching, and assessments. Kenya's teacher education programs can provide a forum for discussion and provisional implementation. Keiyo culture provides but one example of the undocumented potentials of indigenous science education systems. The need for maintaining the richness of cultural diversity, including science education, is as imperative as maintaining species diversity. Science educators now have an opportunity, responsibility, and privilege to become a part of the solution to problems facing the world's indigenous cultures. "Edudiversity" is as important as biodiversity, and "ethnoscience education" is as important as ethnobiology.

References

Ackland, J. (1971). *East African crops*. London: Longman.

Agnew, A. (1974). *Upland Kenya wildflowers*. London: Oxford University Press.

Aikenhead, G. (1994). The social contract of science: Implications for teaching science. In J. Solomon & G. Aikenhead (Eds.), *STS education: International perspectives on reform* (pp. 11–20). New York: Teachers College Press.

BioQUEST. (1998–99). *BioQUEST Library, Volume V*. New York: Academic Press.

Bogonko, S. (1992). *A history of modern education in Kenya (1895–1991)*. Nairobi: Evans Brothers (Kenya) Ltd.

Bray, M., Clarke, P., & Stephens, D. (1986). *Education and society in Africa*. London: Arnold.

Brenzinger, M. (1992). *Language death: Factual and theoretical explorations with special reference to East Africa*. New York: Mouton de Gruyter.

Brunner, J. (1977). *The process of education*. Cambridge: Harvard University Press.

Central Intelligence Agency. (2000). *The world fact book—1999*. Available online: http://www.odci.gov/cia/publications/factbook/ke.html

Chesaina, C. (1991). *Oral literature of the Kalenjin*. Nairobi: East Africa Publishers, Ltd.

Churchill, W. (1962). *My African journey*. London: Neville Spearman Ltd.

Cobern, W. (1996). Constructivism and non-Western science education research. *International Journal of Science Education, 18*, 295–310.

Cox, P. (2000). Will tribal knowledge survive the millennium? *Science*, 287, 44–45.

Darden, L. (1991). *Theory change in science: Strategies from Mendelian genetics*. New York: Oxford University Press.

Dobzhansky, T. (1973). Nothing in biology makes sense except in the light of evolution. *The American Biology Teacher, 35*, 125–129.

Driori, G. (1998). A critical appraisal of science education for economic development. In W. Cobern (Ed.), *Socio-cultural perspectives on science education*. Boston: Kluwer Academic Publishers.

Eshiwani, G. (1993). *Education in Kenya since independence*. Nairobi: East African Educational Publishers.

Finley, F., Stewart, J. & Yarroch, W. (1982). Teacher's perceptions of important and difficult science content. *Science Education, 66*, 531–538.

Freire, P. (1996). *Pedagogy of the oppressed*. New York: Continuum.

Gachathi Report. (1976). *Report of the national committee on educational objectives and policies*. Nairobi: Government Printers.

Gerdes, P. (1995). *Ethnomathematics and education in Africa*. Stockholms Universitet: Institute of International Education.

Gerdes, P. (1999). *Geometry from Africa: Mathematical and educational explorations*. Washington, DC: The Mathematical Association of America.

Glynn, S., Duit, R. & Thiele, R. (1995). Teaching science with analogies: A strategy for constructing knowledge. In: S. Glynn and R. Duit. (Eds.) *Learning science in the schools*. (pp. 247–274). Mahwah, NJ: Lawrence Erlbaum Associates.

Goldschmidt, T. (1996). *Darwin's dreampond*. Cambridge: MIT Press.

Haltenorth, T., & Dillar, H. (1996). *Mammals of Africa including Madagascar*. New York: HarperCollins.

Hess, D. (1995). *Science & technology in a multicultural world.* New York: Columbia University Press.

Jegede, O. (1994). African cultural perspectives and the teaching of science. In J. Solomon & G. Aikenhead (Eds.), *STS Education: International perspectives on reform* (pp. 120–130). New York: Teachers College Press.

Juma, C. (1989). *The gene hunters: Biotechnology and the scramble for seeds.* Princeton, NJ: Princeton University Press.

Kenya Institute of Education. (1992). *Secondary education syllabus: Volume 7.* Nairobi: Kenya Literature Bureau.

Kenya Institute of Education. (1997a). *Secondary biology and biological sciences: Pupils' book one.* Nairobi: Kenya Literature Bureau.

Kenya Institute of Education. (1997b). *Secondary biology and biological sciences: Pupils' book two.* Nairobi: Kenya Literature Bureau.

Kenya Institute of Education. (1997c). *Secondary biology and biological sciences: Pupils' book three.* Nairobi: Kenya Literature Bureau.

Kenya Institute of Education. (1997d). *Secondary biology and biological sciences: Pupils' book four.* Nairobi: Kenya Literature Bureau.

Kokwaro, J. (1976). *Medicinal plants of East Africa.* Nairobi: East African Literature Bureau.

Krauss, M. (1992). The world's languages in crisis. *Language,* 68, 4–10.

Larsen, T. (1996). *The butterflies of Kenya and their natural history.* London: Oxford University Press.

Leakey, R., & Lewin, R. (1992). *Origins reconsidered.* New York: Doubleday.

Linthicum, K., Anyamba, A., Tucker, C., Kelley, P., Myers, M., & Peters, C. (1999). Climate and satellite indicators to forecast Rift Valley Fever epidemics in Kenya. *Science,* 397–400.

Lonsdale, J. (1989). The conquest state, 1895–1904. In W. Ochieng' (Ed.) *A modern history of Kenya 1895–1980* (pp. 6–34). Nairobi: Evans Brothers (Kenya) Ltd.

Mackean, D. (1980). *Introduction to biology: New tropical edition.* London: John Murray.

Makotsi, R. (1996). *Why mosquito bites man.* Nairobi: East African Educational Publishers.

Mazuri, A. (1986). *The Africans: A triple heritage.* London: BBC Publications.

Mbiti, J. (1969). *African religions and philosophy.* Oxford: Heinnemann Educational Publishers.

Moore, J. (1993). *Science as a way of knowing: The foundations of modern biology. Cambridge.* MA: Harvard University Press.

National Research Council. (1996). *Lost crops of Africa.* Washington, DC: National Academy Press.

National Research Council (1996). *National science education standards.* Washington, D.C.: National Academy Press.

Ngugi wa Thiong'o. (1986) *Decolonising the mind: The politics of language in African literature*. Portsmouth, NH: Heinemann.

Njoroge, R., & Bennaars, G. (1994). *Philosophy and education in Africa*. Nairobi: Transafrica.

p'Bitek, O. (1984). *Song of Lawino & Song of Ocol*. Oxford: Heinemann International Literature and Textbooks.

Peters, C. (1891). *New light on dark Africa*. London: Ward, Lock, & Co.

Plotkin, M. (1994). *Tales of a Shammon's Apprentice*. New York: Penquin Books.

Republic of Kenya. (1964). *Kenya education commission report, Part 1*. Nairobi: Government Printer.

Republic of Kenya. (1976). *Report of the national committee on educational objectives and policies*. Nairobi: Government Printer.

Richmond, M. *The seashores of Eastern Africa*. Stockholm: Sida.

Russell, P. (1996). *Genetics*. NY: HarperCollins.

Samoff, J. (1996). *Analyses, agendas, and priorities for education in Africa*. Paris: UNESCO.

Schwab, J. (1962). *The teaching of science*. London: Oxford University Press.

Shiundu, J., & Omulando, S. (1992). *Curriculum theory and practice in Kenya*. Nairobi: Oxford University Press.

Sifuna, D. (1990). *Development of education in Africa: The Kenyan experience*. Nairobi: Initiatives Publishers.

Sommer, G. (1992). A survey of language death in Africa. In M. Brenzinger (ed.) *Language Death: Factual and Theoretical Explorations with Special Reference to East Africa*. (Berlin: Mouton de Gruyter). 301–417.

Soper, R., & Smith, S. (1995). *Biology: An integrated approach for East African Schools*. Nairobi: MacMillan.

Stabler, E. (1969). *Education since Uhuru: The schools of Kenya*. Middletown, CN: Wesleyan University Press.

Stone, R., & Cozens, A. (1977). *New biology for tropical schools*. London: Longman.

Thomson, J. (1887). *Through Maasai land*. London: Frank Cass & Co.

Thomson, N. (1993). *Problem solving in transmission genetics: A new perspective*. Unpublished Ph.D. dissertation. The Graduate School. University of Wisconsin—Madison, WI.

Thomson, N. (2000a, April). *Preservation of cultural knowledge using an indigenous Keiyo game: Historical contexts, cognitive strategies, and connections to problem solving in transmission genetics*. Paper presented at the American Educational Research Association, New Orleans.

Thomson, N. (2000b). Standards are needed, but what should they be? In T. Koballa & D. Tippins (Eds.), *Cases in middle and secondary science education: The promise and dilemmas*. Upper Saddle River, NJ: Merrill.

Thomson, N. & Stewart, J. (1985). Secondary school genetics instruction: making problem solving explicit and meaningful. *Journal of Biological Education, 19*, 53–62.

Timberlake, L. (1986). *Africa in crisis: the causes, the cures of environmental bankruptcy.* Philadelphia: new society publishers.

TIQET. (1999). Totally integrated quality education and training. *Report of the commission of inquiry into the education system of Kenya.* Nairobi: Kenya Government Press.

Tsuma, O. (1998). *Science education in the African context.* Nairobi: The Jomo Kenyatta Foundation.

Wass, M. (1999). *Taking note of language extinction.* [On-line]. Available: *http://www.colorado.edu/lec/alis/articles/langext.html.*

Williams, J., & Arlott, N. (1996). *Birds of East Africa.* New York: Harper-Collins.

Willinsky, J. (1998). The obscured and present meanings of race in science education. In D. Roberts & L. Ostman (Eds.), *Problems of meaning in science curriculum.* New York: Teachers College Press.

Wilson, E. (1996). *Biodiversity.* Washington, DC: National Academy Press.

Wuethrich, B. (2000) Learning the world's languages—before they vanish. *Science, 288,* 1156–1159.

Zaslavsky, C. (1991). *Africa counts: Number and pattern in African culture.* Brooklyn: Lawrence Hill Books.

Zerner, C. (2000). *People, plants, & justice: The politics of nature conservation.* NY: Columbia University Press.

Chapter 9
Culturally and Politically Based Education
A Story of Latin American Science Education Development

WILLIAM R. VEAL, M. FRANCISCA DUSSAILLANT, AND
GIOVANNI OBANDO ROMAN

The topic of science teacher preparation has long been an issue in the United States. There have been many national documents and reports espousing the limitations of and needs for quality science teacher education based upon results on national and international tests. Lately, there has been an increasing interest in how countries would react to the Third International Mathematics and Science Study and the urgent need for science teachers. The renewed interest in science teacher preparation in Latin America is based upon three interrelated premises accepted by most educators in the region. First, the quality of a country's educational system depends on the quality of the teachers. Second, there is interest in the correlation between the professional preparation of teachers and their practices in the classroom. Third, there is a belief that teachers' practices have a significant effect on students' academic performance and learning. This chapter presents issues facing science teacher education in Latin America focusing on case studies from seven countries: Bolivia, Brazil, Chile, Costa Rica, Guatemala, Mexico, and Venezuela.

The purpose of this chapter is to present models of professional science development in several Latin American countries (LAC). Specifically, education in the region is introduced from a cultural and political view, a descriptive background of the countries is presented, educational systems are outlined for each country, teacher preparation programs are delineated, science teacher preparation at all levels is discussed holistically, and science education themes relevant to the region are introduced. Based upon the information compiled and presented in this chapter, the reader should have a broader understanding of LAC educational systems, models of science teacher education, cultural issues facing science education, and contextual factors influencing the teaching and learning of science in the selected countries. It is assumed that a discussion of the models of science teacher preparation would not have meaning if decontextualized. Thus, we chose to present the educational systems of several Latin American countries first so that the teacher preparation models made sense in relation to the level of teaching in the schools. Because of the lack of information available on science teacher education in LAC published in English, a chapter like this one is warranted.

Purported failings in the educational systems of Latin American countries are particularly alarming because economic development is increasingly linked to scientific and technological knowledge. In addition, the political climate influences the school system and ultimately how teachers are trained and certified. Overall, there is now a widespread desire for change, reflected in a wave of reforms in science education that has taken place around the region in the last few years. Just as there are common political and economic similarities, the science education reforms have certain common features.

The common features among the seven Latin American countries presented in this chapter are influenced by economic, cultural, and political factors. The seven countries were specifically chosen for their representation of major issues facing similar countries in the region. Bolivia was chosen because of its position as one of the most rural, poor, and culturally mixed countries, yet it has adopted a tertiary certification program for its science teachers. Brazil was chosen because of its economic stability, regional influence, educational reform issues, and size. Chile was chosen because of its political and economic stability, which has helped it initiate reforms more readily than other countries, such as their science teacher tertiarization within higher education institutions. Costa Rica was chosen because of its political stability and focused approach to teacher education. Guatemala was chosen because it is one of the most rural, poor, and culturally mixed countries in Central America. Mexico was chosen because it is our economic and geographic neighbor and many students in our schools had their initial learning experiences in Mexican schools. Venezuela was chosen because of the political influence past and current presidents have had and will have on the educational system in the country.

The information presented in this chapter was collected over a period of 3 months through reviews of published and available unpublished materials (in Spanish, Portuguese, and English) and interviews with certified teachers, educators working in the Ministries of Education, and professors at teacher education institutions in Latin American countries. The available information on initial teacher training efforts in Latin America is scattered, outdated, and not organized in a format useful for science education. Only recently, since 1995, have there been some documented and concerted efforts to study and report on the status of teacher training in Latin America (e.g., Messina, 1997). Worse yet, there is a lack of material available on science teacher education (e.g., Furio, 1994; Nuñez Jimenez, 1992). Thus, this chapter uses a multitude of knowledge bases to support our description of science teacher education. These knowledge bases include understandings about teacher shortages, cultural norms, science reform initiatives, and educational systems. Together these paint a picture of the science teacher education efforts and initiatives in these countries.

Background of Countries

The seven countries highlighted in this chapter range in population from 4 million to 160 million. All countries share similar historical Iberian beginnings emanating from Spain and Portugal. The predominant religion is Catholicism; the governments are democratic and have executive, legislative, and judicial branches; and the population is a mix of mestizo, African, native, European, and Asian descent. Bolivia has the greatest diversity of cultural and ethnic groups. Bolivia and Guatemala are two of the least-developed countries in Latin America, and about two-thirds of their people, many of whom are subsistence farmers, live in poverty. Brazil and Mexico are two of the world's leading economic players in natural and manufactured goods. Chile and Costa Rica have emerged as the most democratic countries in the region. Venezuela presents a case in which the government has decreed certain initiatives in response to foreign influences and interests. These Latin American countries not only have great diversity but there also exists many similarities based upon their common historical and colonial roots. They share a common history in their language, education, and culture.

The colonial states in Latin America retained the same organizational structure from European society and established a distinct social pyramid. The colonialists regarded education as an important tool for indoctrinating the indigenous people in religion, language, and social status. This effectively marginalized the educational opportunities for the indigenous people. The church, representing the Crown, established schools and universities for the colonialists and developed nonformal educational programs and institutions for the indigenous people.

Considering the common historical and cultural backgrounds of these countries, one might think that their education systems and teacher education programs are similar. To an extent, this is true. Yet each of these countries has developed its own systems and programs to meet the new needs of a growing and urbanizing population. For instance, the Central American countries studied have some type of "middle school," whereas South American countries have no apparent middle level. In some regions and cities of Brazil the elementary grade has been divided into two halves to facilitate decentralization. Another difference is how these countries have outlined the process for certification. Many countries are going away from a normal school model and toward a postsecondary or tertiary program for most levels of teacher education.

Educational Systems

To understand the teacher preparation for these countries, it is necessary to understand the educational systems to equate the licensure programs with the educational levels. The countries' educational systems are structured along the lines of nineteenth-century French and German models. These models emphasize choice and decision making that allow students and/or parents to direct students' educational futures along different tracks. The educational systems are arranged into levels. The primary level is considered compulsory for all students in these Latin American countries. Once the compulsory education ends, students enter a type of secondary school in which they choose between two congruent paths, either liberal (college preparation) or technical (vocational) education. Students graduating from either secondary level may enter the workforce or continue to the university level. Table 9.1 contains a summarized version and comparison of educational systems among the seven countries highlighted in this chapter.

Teacher Preparation

Teacher preparation includes all programs that prepare teachers to teach in a public or private school at the K–12 level. The programs are found at the secondary, tertiary-university, and nonuniversity institutions. Historically, normal schools have left an imprint on the region. Recently there are shifts away from this model toward a tertiary organizational scheme. It is no longer the norm to find initial teacher training programs at a specific level of the educational system. There are three models of teacher training starting with different programs at different educational levels (i.e., secondary and tertiary), continuing with various types of institutions preparing similar teachers (i.e., teacher preparation schools, secondary schools, and universities), and culminating with quality differences of similar institutions (i.e., teacher preparation schools and universities).

Teacher preparation is somewhat different from that in the United States. For example, a person in Brazil, Costa Rica, Mexico, and Guatemala may enter the primary teaching force after secondary schooling. Part of the teacher certification and training still remains at the secondary level. If a person would like to teach grades 1–4 in Brazil, then he or she can enter a teaching high school, called a Magisterio. Other countries require a university degree in order to obtain a certificate to teach at the elementary level. For a secondary position, most teachers enter a teacher training school and major in education (i.e., science, math, social studies). Most university educated

Table 9.1.

Primary and Secondary Educational Systems in Selected Countries in Latin America

Country	School Level	Grade (Age)	Certificate	School Level	Grade (Age)	Certificate	Type of School	Grade (Age)	Certificate
Bolivia	Primary	1–8 (7–14)		Secundaria			General	9–12 (15–19)	Bachillerato en Humanidades
							Technical	9–12 (15–19)	Bachillerato en Humanidades or Técnico Medio
Brazil	Fundamental	1–8 (7–14)	Certificado de Concluso de Primeiro Grau	Ensimo Nivel Medio			General	9–11 (14–17)	Diploma de Segundo Grau
							Profesionalizante	9–12 (14–18) Medio	Diploma de Técnico de Nivel
							Magisterio	9–12 (14–18)	Habilitacao para, or Magisterio de Infantil a 4 serie do 1 grau
Chile	Primary	1–8 (6–13)	Primary School Learning Certificate	Secundaria			Educación Media Humanistico-Cientifico	9–12 (14–18)	Licencia de Educación Media
							Técnico-Profesional	9–13 (14–19)	Técnico de Nivel Medio
Costa Rica	Primary Cycles I & II	1–6 (6–12)		Lower Secondary Cycle III	7–9 (12–15)	Certificado			
				Higher Secondary Cycle IV			Colegio Diversificado	10–11 (15–17)	Bachiller en el Nivel Medio

Table 9.1. (Continued)

Primary and Secondary Educational Systems in Selected Countries in Latin America

Country	School Level	Grade (Age)	Certificate	School Level	Grade (Age)	Certificate	School Level	Type of School	Grade (Age)	Certificate
Guatemala	Primary	1–6 (7–13)		General Secondary	7–9 (13–16)			Colegio Diversificado, Técnico, Profesional	10–12 (15–18)	Técnico en el Nivel Medio or Bachiller en el Nivel Medio
								Diversified Secondary	10–11 (16–18)	Bachillerato
								Upper Secondary	9–11 (15–18)	Perito Industrial
Mexico	Escuela Primaria	1–6 (6–12)	Certificado de Primaria	General Secondary	7–9 (12–16)	Bachillerato General	Higher Secondary	Educacion Media Superior	10–12 (16–18)	Bachillerato, Tecnológico, or Profesional
				Technical Secondary	7–9 (12–15)	Certificado de Secundaria				
Venezuela	Primary	1–8 (6–14)	Certificado de Aprobación				Secondary	Educación Media, Diversificada	9–11 (14–16)	Título de Bachiller
								Educación Media, Profesional	9–12 (14–17)	Técnico Medio

and certified science teachers usually accept positions in the private secondary schools. Table 9.2 summarizes the teacher education processes for licensure in the seven selected Latin American countries.

During the last four decades there has been a reformation of normal schools. The evolution process has decentralized the monopoly on teacher training. The first model of evolution transferred all responsibilities to institutions of higher education (Chile). The second model kept normal schools at the middle level of education and tertiarization has not taken place (Brazil and Guatemala, although Brazil is currently moving away from this). The third model is the tertiarization of normal schools (Bolivia, Costa Rica, Mexico, and Venezuela). Accompanying the three evolutionary models for initial training are four reform models for teacher education in Latin America. The reforms have spurred a transformation of normal schools from middle to a higher nonuniversity level and established the university as another place for certification and training. This transformation from secondary to postsecondary is called tertiarization (Messina, 1996). The four models contribute to teacher training by highlighting the university as the source of knowledge in the region.

Brazil and Guatemala may be considered in the conservative model, in which teacher training at normal schools remains at the secondary level. Brazil trains its elementary teachers at three levels: secondary, normal, and university. Some proponents want to terminate the Magisterios and create additional courses at the university level to replace those at the secondary level. Guatemala, because of its cultural and ethnic diversity and its rural population, has maintained its secondary teacher training at normal schools.

The second model incorporates the tertiarization towards nonuniversity higher education models. Venezuela is the only country of these seven that has established an independent institution for the education of teachers. The Universidad Pedagógica Experimental Libertador certifies teachers at all levels. Students take classes in content areas while also majoring in education (i.e., secondary biology education).

The third model is called the hierarchical model for normal schools. Bolivia and Mexico have transfered normal schools from secondary level to a higher education level by granting them new functions or titles (i.e., higher education normal schools). Mexico has 473 higher education normal schools and 74 units of the Universidad Pedagógica Nacional.

The fourth model is university oriented. Costa Rica and Chile have shifted the entire teacher training process to the university level. In Costa Rica, the University School of Pedagogy coexisted with postsecondary normal schools in the 1960's. In 1973 normal schools transferred to the domain of three universities. In Chile, secondary-level normal schools coexisted with universities for teacher training until 1974 (normal schools were already afforded the higher education level in 1967). Currently, only the universities

Table 9.2.

Licensure Tracks for Teacher Education in Select Latin American Countries

Country	Program Type and Years to Complete	Type of School	Degree Needed to Become Certified	Program Type and Years to Complete	Type of School	Program Type and Years to Complete	Type of School	Degree needed to become Certified
Bolivia	Primaria, 3	Escuelas Normales Superiores	Bachillerato en Humanidades			5	Escuelas Normales Superiores and Universities	
Brazil	Fundamental, 4	Magisterio or Universities	Diploma de Segundo Grau or Licencia			Ensimo Medio, 4	Universities, Teacher Training Institutions	Licenciatura
Chile	Primaria, 3–5	Universities	Licencia Profesorado, Bachillerato Universitario, Licenciatura			Secundaria, 4–5	Universities	Profesor
Costa Rica	Primaria, 2–3 or 4–5	Universities				Secundaria, 4–5	Universities	Bachillerato Profesorado, Licenciatura
Guatemala	Primaria, 3	Teacher Training Colleges	Bachillerato			Secundaria, 4	Universities	Profesor
Mexico	Escuela Primaria, 4	Escuela Normal	Licenciatura	Lower Secundaria, 4	Escuela Normal and Universities	Higher Secundaria, 4	Escuela Normal and Universities	Licenciatura
Venezuela	Primaria, 4	Teacher Training Institutions, Pedagogical Institutes, Universities	Licenciatura			Secundaria, 4	Teacher Training Institutions and Universities	Licenciatura, Título Profesional

by law are allowed to partake in teacher training. There are nonuniversity institutes that also do teacher training, but these exist because of teacher shortages and are not regulated by the government.

Because of the differences in educational structures within the countries, it is difficult to state a norm for teacher education in the region. Most of these institutes (i.e., normal schools and universities) are teaching centers with little or no research or extension activities. The recent reform initiatives indicate a movement to include more research-based ideas and processes that will soon be part of the norm at these institutions. Some countries, like Bolivia, Costa Rica, and Mexico, integrate inservice education with their institutional mandate.

There exist important differences within countries because of teacher demand, economic situation, and level of certification. Additional differences exist within countries because not all of the training imparted by the universities is equivalent. For example, the number of years to train a teacher fluctuates between 2 and 5. In another example, the normal schools in five of the seven countries highlighted do not have Web sites. The normal school in Venezuela does have a Web site, but no programmatic information could be obtained. The tertiary universities were the only programs that had information available, and sometimes the sites were poor and contained very little information. This is interesting in light of the significance Venezuela places on the integration of technology and Internet use in science teacher preparation. Even though the normal schools may have more traditional teacher preparation, the tertiary universities are better equipped to introduce current reform and technological initiatives. Other programmatic similarities are found across the countries. The following section highlights some of these similarities.

Science Teacher Preparation

The state of science teacher education has recently become a topic of discussion in some Latin America countries because of the disjointed knowledge used in the preparation of science teachers (Cudmani & Pesa, 1997; Davini, 1995). The debate focuses on mastering the subject matter, eradicating the simplistic view of the nature of science, developing new instructional strategies, and linking science teaching and science educational research. This gap between the scientific discipline and pedagogical formation is similar to the current debate in the United States centering on pedagogical content knowledge (McDermott, 1990; Shulman, 1986, 1987). This fragmentation can be seen in who teaches and how many courses a student takes in the methods or didactics course in science. For example, Nuñez-Jimenez (1992) determined that the percentage of hours in a degree program that is devoted to science methods (didactics) is only 5 percent and 7 percent for Chile and Brazil, respectively. Another aspect that confounds the interpretation and ability to compare programs among universities and countries is the definition and the use of terms such as *didáctica, metodología,* and *pedagogía.* All three words mean "methods," but they are used in different contexts in different countries. For example, *didáctica* means general methods in Bolivia and science specific methods in Venezuela, *metodología* means science specific methods in Chile, and *pedagogía* means general methods in Venezuela and science specific methods in Mexico.

Primary

At the elementary or primary level, there may be anywhere from one to three content/methods courses. Students in these classes learn not only the content but also the methodologies for teaching that content. In some countries teachers need a secondary degree or three years of postsecondary education for a license. The programs mostly consist of education classes ranging from the psychology of education and didactics to the history of education (Domingues, Koff, & Moraes, 1998). Thus, the content knowledge is very limiting. Teachers take few science and science pedagogy courses. On the other hand, many students may have adequate content background because of their secondary schooling and preparation to enter a university.

For those countries that have middle-level schools or a divided elementary school, science teachers need to have a strong background in content to teach. For example, in Brazil a science teacher of grades 5–8 in the Infantil level needs a biology license (Licenciatura en Ciencias Biologicas). Only in areas in which there is a shortage of teachers are general elementary teachers teaching science at the middle level without a strong background in science content.

Secondary

At the secondary level, most certified teachers enter college with science teaching as a major, rather than a specific domain of science as a major. These students will predominately enter a teaching institution, normal school, or university and receive training in a specific domain of science and its pedagogy. At least in one university (Universidad de los Andes, Venezuela) a person may enter as a science education major, but decide upon a specific domain of science in his or her third year. Science teachers are competent in their content knowledge and take science methods classes in their domain of science. A professor who has a strong background in both science and education usually teaches these classes. Historically, science content specialists taught the methods classes. The field of content-specific education is advancing in Latin America and more professors are specializing in science education. There was only one exception found to the predetermined major of science education. In the Pontificia Universidad Católica de Chile, students who have degrees in a science discipline can take an additional year of pedagogy courses to become certified. Table 9.3 highlights some of the characteristics that different South American universities have for obtaining a secondary science certificate.

Costa Rica has a unique type of program and system for certifying science teachers. All three universities that license science teachers follow the same program of study. Science teachers may obtain three different types of licenses depending upon the number of years of study: three for the Profesorado, four for the Bachillerato, and five for the Licenciatura. Table 9.3 shows the number of science courses required for the different degree levels in Costa Rica. Those who obtain the Licenciatura usually attain employment at the best private schools. Currently, anyone can teach in the public schools with 1 or 2 years of higher education and without a degree. The licenses also differ in the number of science and education classes taken. One result of this differentiation is that teachers get paid differently based upon the level of certification. Another distinction is that teachers holding the Bachillerato or Profesorado license may teach science only in the "middle" (Cycle III) or secondary (Cycle IV) school. Those teachers who continue their studies and obtain a Licenciatura may teach at any level including the university. Finally, Costa Rica is the only country that offers a general science certification. Science teachers are no longer awarded licenses to teach in one domain of science. This is a recent change in policy in the past 6 years.

There are many similarities between countries and programs in LAC. As shown in Table 9.3, most science teachers at the secondary level become certified in two fields, chemistry and biology or math and physics. This dual certification is prevalent at the tertiary levels. Compared to the United States, geology or earth science in LAC is found under the auspices of geography. This includes the study of Earth systems, atmosphere, tectonic processes, and mapping. Most of this content can be found at the primary level rather than the secondary level of public schooling and certification. Environmental science is found within biology.

The practicum or student teaching internship varies among universities and countries. Villegas-Reimers (1998) stated that most teachers have an adequate preparation in the content area and pedagogy, but lack the practical aspects of teaching. There are few set national standards or guidelines. For example, Chile requires by law that preservice teachers teach one semester as an intern before becoming certified. The Universidad de Serena in Chile requires students to take two courses in two semesters in which they observe classes, two courses in two semesters in which they "help" a classroom

Table 9.3.
Sample Secondary Science Education Program Descriptions

	Bolivia	Chile	Chile	Venezuela	Venezuela
University	Universidad Católica Boliviana	Universidad de la Serena	Pontificia Universidad Católica[a]	Universidad de los Andes	Universidad Católica
License Area	Biology or Chemistry	Chemistry and Natural Sciences, Biology and Natural Sciences, or Physics and Math	Natural Sciences and Biology, Chemistry, or Math and Physics	Biology, Chemistry, or Physics	Biologic Sciences or Math and Physics
Years of Study	5	4.5	1	4.5	5
Number of General Methods Courses[b]	2	0	1	1	1
Number of Science Specific Methods Courses[c]	2[d]	2—Chemistry and Biology, 3—Physics and Math	1[d]	2[d]	1[d]
Number of Science Courses[e,f]	19	13—Profesorado, 17—Bachillerato, ?—Licenciatura	13—Profesorado, 17—Bachillerato, ?—Licenciatura	8—Biology, 12—Chemistry and Biology	18—Biology, 11—Physics, and Math

[a] This is a one-year teacher certification program. As an entry requirement, the student must hold a bachelor's degree in a science-related field.

[b] These classes have titles under the names Didáctica, Pedagogía, and Metodología.

[c] Some examples of course titles are: Metodología de la Enseñanza de las Ciencias Naturales, Metodología de la Enseñanza de la Química, Didáctica de las Ciencias Físico, Epistemiología y Procesos de la Enseñanza-Aprendizaje de la Física Biología, and Recursos para el Aprendizaje de la Química.

[d] The methods course is specific to the science content area or domain.

[e] The math content courses that everyone should take for their particular certification area are not included. Those in chemistry and physics are required to take more math courses, thus they take fewer science courses.

[f] In Chile, students may obtain a Profesorado, Bachillerato, or Licenciatura. The difference is in the number of science content courses and education courses. The courses for the Licenciatura year are currently being revised.

teacher, and a final semester that they devote completely to student teaching. Other countries do not have specific rules or regulations.

Themes Found in Latin American Science Education

As mentioned earlier, there are many science education reform initiatives in LAC. These reform initiatives are driven by different influences. One theme that is not discussed in this chapter is the importance of results from the Third International Mathematics and Science Studies (TIMSS). Part of the reason for relatively no LAC participating in these studies is the fear of being compared to other countries. As it is, Colombia was the only country that competed in the Second International Mathematics and Science Study. Their performance was so low compared to the world that other countries in the region were fearful of repeating the same embarrassment. As a way of evaluating its educational quality, Chile is the only LAC to have participated in the 2000 TIMSS. Rather than discuss TIMSS and its implications for science teacher education in these countries, we will focus on assessment and evaluation in these countries. Other themes are obvious when comparing these countries to the United States. One is the difference between public and private education. This distinction takes on a new meaning in the Latin American culture. A third theme is the type of reform initiatives that these countries are trying to initiate at multiple levels. The reform initiatives are focused at curriculum and teaching that ultimately has implications for training teachers to be ready for those changes in the schools. The final theme is the political and economic realities of each country that drives the decisions for funding, assessment, and curriculum.

Evaluation

One of the primary objectives of the modernization of the Chilean education system was to improve the quality of education. To measure this progress the System for Measuring Educational Quality (SIMCE) was devised to establish a quantitative appraisal of the behavior of certain factors linked structurally to the quality of the educational services provided. The SIMCE was first implemented in 1988 and consisted of a series of tests aimed to ascertain the cognitive knowledge and affective area development of around 85 percent of students in fourth, eighth, and eleventh grade. The areas measured are Spanish, mathematics, and social and natural sciences.

Mexico has developed two national assessment examinations to measure student progress. The Programa de Carrera Magisterial is used to evaluate the professional performance of teachers, in which standardized tests are used to measure student progress. The Programa de Instalación y Fortalecimiento de las Areas Estatales de Evaluación is a test at the primary level to assess the national standards for different subjects and grades of basic education. These evaluations have served to inform people on accreditation, selection, and promotion of teachers and pupils. No test at the secondary level has been created yet.

Many countries have tests for college entrance. In some countries, like Brazil and Chile, these college entrance exams are very competitive, but entrance into a teaching program is relatively easy and not very competitive. This information is important in that it informs us of the science content preparation of many primary and secondary school teachers. Since many primary school teachers in Latin America do not take content courses at the university or normal school it is important to know that their science content background may be poor because of their secondary education.

Private versus Public Schools

The main distinction between the educational system in Latin America and the United States is the differentiation between private and public schools. The establishment and existence of the two types reflect the social economic situation within the countries. People are generally categorized into two levels, those with money and those without. What is meant by private school often differs between Central and South American countries. In many Central American countries, community schools are classified as private and are created to cater to rural educational needs. Liang (1999) reported that the percentage for teachers in private schools in certain Latin American countries ranges from 9 percent in Costa Rica to 39 percent in Chile. In Brazil, the percentage is 31 percent, but divisions within the private schools exists. There are private schools with some money and support to sustain a growing middle class and others with lots of money and support. Private schools are considered the best schools for teaching and learning. They are usually funded by private fees, and are thus reserved for wealthy families. In some instances, private schools receive vouchers from the government. Various studies from Chile on the effectiveness of the voucher system on student achievement on the SICME exams are inconclusive. Based on the preceding information, the best science instruction and learning occurs, presumably, in the privately funded schools. This is deduced based upon higher salaries, money for supplies, and access to technology to supplement science instruction.

Pedagogical experience and training is usually a luxury reserved for the private schools. In many countries the requirement for pedagogical training, because of the teacher shortage, has been dropped or lessened. Thus those teachers with solid backgrounds in a science discipline will teach in private schools. Those with a science and pedagogy background will also teach in private schools. This also has implications for student achievement between rural and urban, poor and nonpoor areas. Nonpoor students enrolled in urban schools have more educational opportunities than their rural and urban poor peers. The disparities in access to quality education were seen in a 1992 pilot study of TIMSS on 13-year-olds. Students who attended elite private schools outperformed those in rural public schools on the science portion of the test by 20 percent in Costa Rica and 25 percent in Venezuela (World Bank, 2000).

Reform Initiatives

The rationale behind many of the current reform issues stems from societal and economic reasons. In all developing and industrial countries there is a push to make citizens more science and technologically literate. Despite the common goal, there is considerable disagreement on the means (Ware, 2000). Many countries are implementing reforms to meet the common goal, many of which will have a profound impact on the training of science teachers at all levels. Garritz and Talanquer (1999) mentioned that science education reform "will be considered one of the greatest transformations of Mexican education in the 20th century" (p. 76). Martinez and Cerón (1999) stated that the Chilean reform of science education takes place within the wider context of systemic reform, designed to decentralize curriculum decision making and ensure quality by establishing specific mandatory objectives for all students.

Many of the science teaching reforms mirror those of the United States. For example, teachers are trying to integrate science into the local context. Using problems that affect the community, teachers endeavor to show the practical value of scientific knowledge in determining the causes of specific phenomena. They encourage students to come up with ways of possibly preventing environmental catastrophes. This reform initiative is similar to the Science-Technology-Society curriculum established in the 1980s. Chem-Comm, developed by the American Chemical Society, has recently pushed a Latin-based curriculum similar to the popular hands-on textbook used in the United States. The focus is on contextual chemistry and processes for solving problems at the local levels. Some sample issues are water quality, mineral development, and petroleum refining.

Another reform example is the Globo project in Costa Rica, the goal of which is to make students aware of environmental protection by studying the El Niño climatic phenomena. Costa Rica has also established the Administration of Environmental Education and Sustainable Development to further the cause of environmental and contextual science education. Doryan and Badilla (1999) described a program emphasizing environmental ethics that was designed to maximize the human potential and develop the economy, while respecting the natural environment. For both of these programs to work, science teacher education will have to integrate proper training of pre-service teachers. This training will have to focus on inquiry and hands-on methodology. Without proper initial training, science teachers will continue to teach traditionally (Ware, 2000).

In Mexico, the 1993 reform in the natural sciences required the replacement of the traditional education model, in which teaching was based on the use of the scientific method as a process of discovery and content was based on the idea of "learning science by doing science" (Talanquer, 1990). In contrast, the new curriculum uses constructivist learning principles. For example, Garritz and Talanquer (1999) wrote that there is a strong emphasis on Science-Technology-Society issues in junior high schools and an increasing emphasis at the secondary level, even though there is no national curriculum at the secondary level. Teachers have been asked to incorporate students' ideas, background knowledge, and experiences into the science discussions. This requires that the teacher training institutions develop proper inquiry-based teacher training for both preservice and inservice science teachers (Garritz & Talanquer, 1999).

Another reform initiative focuses on the curriculum. At the higher primary and lower secondary levels, curricula include more integrated, combined, coordinated, and modular approaches to science. These ideas are drawn from traditional science discipline content and stress scientific skills and cognitive processes above content. These are familiar ideas to many science educators, but in the context of developing countries, there needs to be an emphasis on teacher training. Many of the current science teachers (primary and secondary) in Latin America were taught under the old school of thought that was void of inquiry and constructivist principles. For example, in Chile the inquiry approach to teaching science is starting to gain importance (Martinez & Cerón, 1999).

Political and Economic Factors

The almost universal transition from dictatorship to participatory democracy in the region and the increasingly active civil society have placed new burdens on the educational systems, both for schools and teacher training. For example, teacher shortage is often a result of the country's economic and political situation. For Latin American countries, the need for teachers is a national dilemma that requires attention. In many developing countries there is a direct link between science education and the labor market and economy (Lewin, 2000). Investment in science teacher education that ultimately leads to increased science and technology knowledge in the schools is necessary but not sufficient for economic development. Even though most students will not become scientists, science knowledge should meet the needs of the broader population. The current reform initiatives tied to the economic well-being of a country are allowing science education to grow. Some countries have established inservice training for teachers to become updated in the areas of science and technology (e.g., Bolivia, Chile, Mexico). In particular, Mexico has established inservice training through distance education technology.

One disadvantage as a result of this situation is the lax standards for hiring teachers at all levels. There is a "lateral entry" or plain "hire and teach" mentality in many poor and rural areas. Because of teacher shortages, many teachers at all levels are hired with a basic level, a secondary level, or a university level of education without any courses in education or content-specific methods. The problem is compounded when private schools are considered the best schools in which to teach. Many qualified and noncertified teachers can be found in the private schools rather than in the public schools. Currently, the disparity between public and private schools can be seen in the background and training of teachers, level of students, quality of buildings, availability of supplies, and level of pay.

Venezuela's political and economic situations act as catalysts for educational reform. For example, the newly elected president wants to privatize schools and choose an ally who will oversee the schools. His rationale is to include military education in all schools and take out certain curricular programs such as English and computers. Ultimately, he wants to indoctrinate the students with his revolutionary ideologies. How this will affect the science education reform initiatives and science teacher training is yet to be seen, albeit it is important to note that the political environment in many Latin American countries is more directly related to and influential of the educational atmosphere.

Concluding Remarks

Most people working in education for developing countries agree that science education is an important element for development. In reality, however, science education is perceived as an elite subject that lacks relevance to the social and economic reality surrounding student life and does not reach a large student population. There are efforts to establish more communication and collaboration among science educators in the region. For example, most of the papers presented at two meetings of the Ibero-American Science Education Group focused on a constructivist approach to research, either explicitly or implicitly (Ryan, 1999). The purpose of this group is to develop the practice of professors and teachers as it relates to their students' learning of science. More research-based articles are appearing in science education journals of the region (e.g., *Enseñanza de las Ciencias* and *La Revista Iberoamericana de Educación*).

In all cases we need to be careful not to generalize among these nations. The countries and region covered in this chapter are vast. There are common ideas and conclusions that can be learned from the region as a whole. Inequalities for school access, readiness, and attendance still pervade in the region. The rural poor and indigenous populations are at an extreme disadvantage relative to other groups in terms of the quality of science education and science teachers. The World Bank (2000) stated that most research on quality of schooling and achievement levels in LAC points to a low quality of teaching. For science in particular, this is disturbing. There are signs of closing the gap between the "haves" and "have-nots." For example, Guatemala established Eduquemos a la Niña, which incorporates innovative techniques to expand educational opportunities to girls and rural children and to improve science teaching and learning.

The problems and challenges of teacher preparation faced by the individual countries of Latin America are similar, but they vary depending upon economic and political situations. Science education in developing countries often relies too much on memorization of facts and not enough on learning to understand the relevance of knowledge and its application in the local context. In the industrialized countries there is much greater emphasis on the problem-solving approach, decision making, and developing the ability to analyze and work in a team. Currently, reform movements in science education (similar to those in the United States) represent a huge qualitative change for many countries and directly affect teacher preparation.

The quality of teaching needs to be improved. The reform initiatives are effective only when they are directed at the classroom teachers and students. Many of the initiatives are concentrated in the urban nonpoor and private schools and do not find their way to the poor, public, and rural areas. Many teachers do not provide stimulating or inquiry-based instruction and were never trained in constructivist ideologies that encourage active learning, discussion, and group work. "Evidence from Guatemala suggests that this seems to affect girls in particular, since teachers initiate classroom interventions more with boys than girls, and that boys generally participate more actively in class" (World Bank, 2000, p. 42). A recent study in Brazil confirmed the same results of the quality of teaching between rural and poor and urban and non-poor (World Bank & UNICEF, 1997).

Initiatives that foster individual and organizational incentives are needed to improve science education. There should be two social goals that drive government investment in science education: (1) providing a competent and skilled workforce with science and technology knowledge and (2) reducing social inequality and poverty related to science instruction, access, and materials. Disparities in science educational development and reform exist across and within countries. If the LAC continue to successfully implement various science education reforms, the quality of science teacher education will improve exponentially.

References

Cudmani, L., & Pesa, M. A. (1997). Knowledge integration in the training of physics teachers. *BULLETIN, 43,* 42–48.

Davini, M.C. (1995). *La formación docente en cuestión: Política y pedagogía.* Buenos Aires, Argentina: Paidos.

Domingues, J.L., Koff, E.D., & Moraes, I.J. (1998). Anotacoes de leitura dos parametros nacionais do currículo de ciencias. In E.S.S. Barretto (Ed.), *Os Currículos do Ensino Fundamental para as Escolas Brasileiras.* Sao Paulo, Brazil: Editora utores Associados, pp 193–200

Doryan, E., & Badilla, E. (1999). On the road to improving the quality of life: Environmental education in the Costa Rican education system. In S. A. Ware (Ed.), *Science and environmental education: Views from developing countries.* Washington, DC: World Bank, Human Development Network, Education Group, pp 35–46

Furio, C. (1994). Tendencias actuales en la formación del profesorado de ciencias. *Enseñanza de las Ciencias, 12*(2), 37–44.

Garritz, A., & Talanquer, V. (1999). Advances and obstacles to the reform of science education in secondary schools in Mexico. In S.A. Ware (Ed.), *Science and environmental education: Views from developing countries.* Washington, DC: World Bank, Human Development Network, Education Group, pp 75–92

Lewin, K. M. (2000). Mapping science education policy in developing countries. Washington, DC: The World Bank, pp 1–29

Liang, X. (1999). *Teacher pay in 12 Latin American countries: How does teacher pay compare to other professions, what determines teacher pay, and who are the teachers?* The World Bank Human Development Department: Latin America and the Caribbean Social and Human Development. Paper Series, No. 49, pp 1–33.

Martinez, M., & Cerón, R. (1999). Chilean education reforms during the current century. In S.A. Ware (Ed.), *Science and environmental education: Views from developing countries.* Washington, DC: World Bank, Human Development Network, Education Group, pp 93–104.

McDermott, L. (1990). A perspective on teacher preparation in physics and other sciences: The need of special science courses for teachers. *American Journal of Physics, 58*(8), 734–742.

Messina, G. (1997). How are teachers trained in Latin America. *BULLETIN, 43,* 52–71.

Nunez-Jimenez, S. (1992). *Diagnóstico sobre formación inicial y permanente del profesorado de ciencias y matemática (nivel medio) en los países iberoamericanos.* Chile: Ministry of Education.

Ryan, C. (1999). Developments in science education in Latin America. *Science Teacher Education, 25,* 10–12.

Shulman, L.S. (1986). Those who understand: Knowledge growth in teaching. *Educational Researcher, 15*(2), 4–14.

Shulman, L. (1987). Knowledge and teaching: Foundations of the new reform. *Harvard Educational Review, 57,* 1–22.

Talanquer, V. (1990). Qué pasa en nuestra secundaria? *Educación Química, 1*(2), 92.

Villegas-Reimers, E. (1998). The preparation of teachers in Latin America: challenges and trends. The World Bank Human Development Department: LCSHD Paper Series, No. 15. pp 1–40

Ware, S.A. (2000). Overview of the reform agenda. In S.A. Ware (Ed.), *Science and environmental education: Views from developing countries.* Washington, DC: World Bank, Human Development Network, Education Group.

World Bank. (2000). *Educational change in Latin America and the Caribbean.* Washington, DC: The World Bank, Latin America and the Caribbean Social and Human Development, pp 1–110.

World Bank and UNICEF. (1997). *Programa de pesquisa e operacionalizacao de politicas educacionais. (A call to action, combating school failure in the northeast of brazil.)* Brasilia, Brazil: Author.

Chapter 10

The Role of Onservice Professional Development of Science Teachers in Large-Scale Changes in Teaching Practice
A Look at Britain and Indonesia

PHILIP ADEY

Introduction

David Labaree (1998) suggests that one characteristic of educational research is that it "deal(s) with some aspect of human behaviour. This means that cause only becomes effect through the medium of willful human action, which introduces a large and unruly error term into any predictive equation. These billiard balls are likely to change direction between the cue ball and the pocket" (p. 5). This goes a long way to explaining why, although there has been a century of work attempting to show how the fundamental work of psychologists, philosophers, and sociologists can be translated into classroom procedures that improve the quality of learning, we are still so far from achieving reliable and large-scale recipes for bringing about systemic educational change. We simply don't have strong enough messages from our research. But there are other contributory reasons. Research has too often been limited to lab school experiments conducted at a few sites over a limited time period, with too little promise of generalizability. Blame for this failure may be placed on researchers concerned with academic publication and moving on to the next new work or on their failure as publicists; the conservatism of teachers and school managers; short-sightedness of politicians concerned more with elections a couple of years away than with benefits to society, which may not show up for 10 years; or, inevitably, lack of funding. At different times and in different places probably all of these accusations can be justified and may be summarized as the inertia of the system, like a massive tanker whose direction cannot be changed by a few speedboats of research.

In this chapter I describe two related but remote examples in which a speedboat, by persistence over many years and attention to publicity and multiplier effects, has systematically shifted a tanker of practice. In particular the chapter focuses on the multiplier effect of a professional development program for teachers and for trainers in which *onservice* work (when a trainer goes to the teacher's classroom and coaches the teacher as he or she introduces new teaching methods with his or her own pupils) plays a critical role. For 20 years now, in Indonesia and in the United Kingdom, we have been working with an "inservice-onservice" model of professional development of science teachers which proves to have a powerful impact on large numbers of schools. Sure it has problems of its own, and we need to face these as well as its successes.

I look at the theoretical and empirical underpinning of coaching as a method of bringing about real change in practice and describe more fully the form of the inservice-onservice model. I also offer some evidence from these projects that supports the critical role that onservice coaching plays in effective professional development programs.

Theoretical Framework: Three Strands

In their conclusion to a recent review of research on professional development, Wilson and Berne (1999) wrote, "Our review of the literature leads us to conclude that . . . few such projects had yet completed analyses of what professional knowledge was acquired. . . . Fewer still had explicated their theories of how teachers learned and designed research to test those theories" (p. 204). This may be a rather harsh judgment. I suggest that there are at least three strands of theorizing that have proved fruitful in understanding the process of professional change of teachers. The first is the application of general theories of conceptual or attitude changes to the beliefs and behaviors of teachers; the second is the notion of the reflective practitioner; and the third is an emerging body of scholarship that characterizes teachers' procedural knowledge as "intuitive". Let us consider each of these strands in turn.

Conceptual Change

An example of approaching professional development in the context of conceptual change is provided by Mevarech (1995), who discusses the role of teachers' prior conceptions of the nature of learning and describes the U-shaped learning curve that teachers encounter when trying to replace one skill, and the epistemology on which it is based, with another. The value of this conceptualization of teacher change is that it can draw on the extensive parallel literature on conceptual change and attitude change in students. It leads us to focus on teachers' prior conceptions and to recognize that we are unlikely to bring about change in practice unless we face up to and, if necessary, challenge teachers' deep-rooted beliefs about the nature of knowledge transmission. It suggests the idea that teachers should be included in the process of needs analysis and program design (Joyce & Showers, 1980). It also indicates that such change is likely to be a slow and difficult process and that real change in practice will not arise from short programs of instruction, especially when those programs take place in a center removed from the teacher's own classroom.

Note that in focusing on the need to tackle fundamental concepts and attitudes, we are not necessarily prescribing that this is the first thing that must happen, before change in teaching practice can occur. Indeed, Guskey (1986) has argued persuasively that changes in teachers' beliefs and attitudes may well follow the change in perceived pupil responses, which come about from changed teaching practice. But whether a precursor or a consequence, such deep-seated changes are necessary for permanent effects on teacher practice.

Reflection on Practice

The idea of the teacher as a reflective practitioner has had a long and respectable history in the literature. For example, Baird, Fensham, Gunstone, and White (1991) have shown, particularly with respect to the professional development of science teachers, the central role that reflection—both on classroom practice and on the phenomena of science teaching and learning—has in the pedagogical development process of both preservice student teachers and experienced teachers in inservice courses. More recently, Cooper and Boyd (1999) have described a scheme of peer- and group-oriented reflection on practice developed among teachers in a New York City school district, which provided a systemic self-help strategy for the long-term maintenance of innovative methods in classrooms.

The notion of the reflective practitioner arising from such studies guides practice in professional development by highlighting the importance of allowing teachers time to discuss their current practice and their attempts at changing it during the course of a program. This may be done through diaries or other forms of logs or orally at "feedback" sessions with colleagues and course leaders. In fact the use of the term "feedback" for this process may be somewhat misleading. Feedback per se usually means providing an account of the success or otherwise of a new procedure or activity with an implication that it is the course developer who benefits and is able to modify future advice and developing activities accordingly. On the other hand, in feedback *as reflection*, it is specifically the participating teachers who benefit through putting their experiences and associated feelings into words and discussing them with peers. Experienced course organizers know that such sessions can become self-sustaining and that many of the ostensible "questions" that arise are in fact answered by participants themselves or even serve a rhetorical function: Asking the question provides its own answer. We might ask then whether such self-driven reflection is by itself sufficient to create change in practice—in other words, is there further input required from course leaders who have deeper academic insights into either or both subject content matter or learning theory? Avgitidou (1997), Calderhead (1993), and Korthagen and Kessels (1999) argue from various perspectives that the subject matter of reflection must include, *inter alia*, a richer understanding of the psychological principles of teaching and learning. This gives the leaders of a course of professional development a responsibility to support participant teachers in building such an understanding for themselves and indicates that reflection alone will generally be inadequate to generate change without some expert input.

Intuitive Knowledge in Teaching Practice

Turning now to the intuitive nature of much of the procedural knowledge of teachers, it is important not to confuse the ideas of "intuitive" and "instinctive." The latter implies something in-built, perhaps a personality factor over which no normal professional development course could be expected to have much influence. "Intuitive," on the other hand, implies a behavior that occurs without explicit cognition at the moment at which it arises. The basis of the behavior remains in the unconscious. The term "implicit knowledge" is used (Tomlinson, 1998) for this type of unconscious understanding, which gives rise to intuitive behavior. Intuition is how, as teachers, we react almost instantaneously to situations as they arise in the classroom. A class of 30 or more students is an extremely complex social environment in terms of the quantity and quality of social interactions, which occur minute by minute. There can be rapid changes in mood and style of interaction between the teacher and individual and groups of students. It would be practically impossible for a teacher to proceed in such a situation entirely on the basis of rational and conscious decision making or problem solving. The "professional" response in such situations depends much on intuition, a process well described by Brown and Coles (2000) through their collaboration of a researcher with a practicing teacher. The important point here is that this intuitive behavior is based on our implicit knowledge, and that knowledge is based on previous situations and on the constructs we have built on such experience but not necessarily externalized or made conscious: "[I]t is now well accepted that expert practitioners possess a complex personal knowledge base which they draw upon intuitively. This knowledge base is acquired through training and experience but individuals may not be able to articulate why they do what they do" (McMahon, 2000, p. 138). Such implicit knowledge may be an influence for good or for ill in the direction it proposes for action. Implicit knowledge may be derived from working alongside older colleagues; it may be rooted in an authoritarian view of teacher-student relationships and on a simple transmission model of knowledge transfer and be very traditional in form. On the other hand, it may be derived from a combination of a personal philosophy of guided democracy, with some experience of the process of constructivism, and the observation of colleagues who have shown how all students can be encouraged to contribute to the construction of their own understandings.

The challenge for providers of professional development programs is to devise programs that shape teachers' intuitive understandings such that they improve the quality of daily classroom reactions. Such a process is necessarily slow and cannot follow a simple mechanistic path.

The Three Strands and Onservice Coaching

The three strands of thought on the nature of professional development outlined here are emphatically not alternatives. On the contrary, they may be seen as intertwining and feeding into one another. What is an effective way of inducing a process of conceptual change in a teacher? Why, to encourage reflection. And what is the basis of the intuitive knowledge that guides action? It is the underlying conceptions and attitudes of the individual. Guided reflection assists the process of conceptual change, and conceptual change restructures the intuitive knowledge upon which teaching practice rests. In his seminal work on professional development, Schon (1987) shows how reflection is an essential part of the process by which teachers incorporate the perceived needs of a situation within their own system of beliefs, and this is all part of the development of their "professional artistry," a good description of practice arising from implicit understandings.

It has already been emphasized that each part of the integrated process is itself necessarily slow and difficult, but it can also be shown that onservice work (in-class coaching) relates directly to these three interrelated ways of conceptualizing teacher change. First, the expert coach can provide a mirror to assist the teacher reflect on her or his current practice. A knowledgeable and sensitive observer can feed back to the teacher detail of actual practice of which the teacher (because the practice is largely intuitive) may not himself or herself be aware. This detail in turn provides the raw material for reflection. Either at the time of the lesson or subsequently, either alone or with others, the teacher may then be encouraged to examine her own assumptions, beliefs, and concepts concerning teaching and learning that give rise to her current practices.

A series of such coaching sessions, which may also include occasions when the coach models particular techniques (such as open questioning, wait time, generating activity from all students), promotes a process of conceptual change in the teacher, and the implicit knowledge thus modified gradually changes her behavior. Initially, the new practices may be at a conscious, nonautomated level, when they may be expected to be somewhat stilted and labored, but as they become automated and part of the teacher's intuitive practice so they become fluent and more effective. This process explains the U-shaped learning curve described by Mevarech (1995) referred to earlier, and accords well with our experience of running long-term professional development programs for teachers, which will be described later.

Empirical Support

Before we move to such specifics, some general empirical support for this theory of coaching can be offered. Joyce and Showers (1995) conducted a meta-analysis of studies that review the effectiveness of staff development programs. They used a stringent criterion of "effective," accepting only studies that showed significant effects on the learning of students. In other words, studies that only reported changes in teaching practice without concomitant increased student learning were rejected as "not proven" to be effective. With this criterion, it became very clear that the only staff development programs that were effective were those that included an element of coaching, defined as work with teachers in their own classrooms, an essential supplement to center-based course elements.

To say that coaching is an essential element of effective professional development of teachers is not necessarily to endorse the theoretical constructs of conceptual change, reflection, and intuitive practice outlined earlier. It may be proposed that coaching may also operate in a more direct way by, for example, providing models for the teacher to copy or by showing that the new techniques can indeed work in his classroom with his students. However, all that we know about teacher change suggest that such direct, surface level influences have little permanent effect on teaching practice. Joyce and Weil (1986) found that substantial permanent change in practice requires at least 30 hours of training. We are not looking here at the teacher who suddenly "sees the light" and changes practice overnight. The literature is not replete with examples of such "Road to Damascus" conversions—but see Raw (2000) for an account by a teacher who has seen the light and talks evangelically about teaching for higher-order thinking, but only after a 2-year inservice-onservice professional development (PD) course! We are considering the normal process of hard work over many weeks or months, and I suggest that the only plausible mechanism by which such slow but permanent change could occur is a deep-seated one, well characterized by the three strands of conceptual change, reflection on practice, and intuitive action.

I hope that the discussion so far establishes both the theoretical and empirical justification for including an onservice element in any program for the professional development of teachers. I turn now to two specific examples of such programs to illustrate the practical problems and possibilities involved in providing onservice coaching on a large scale.

The Inservice-Onservice Model in Indonesia

The first of the two projects is the PKG inservice-onservice project in Indonesia, supported by the government of Indonesia since 1980, together with various other funders such as United Nations Educational Scientific &

Cultural Organization and United Nations Development Programme in the early years, and more recently by the World Bank. PKG stands for *Pemantapan Kerja Guru*—improving the work of the teachers. Since its inception it has had an ambitious objective: to reach into secondary schools throughout the 27 provinces of Indonesia, an archipelago of five great land masses and many thousands of islands stretching 5,000 kilometers from West to East and with a population of around 200 million, with the aim of improving the quality of teaching, starting with science and mathematics teaching. From the start, the originators of the project—Dr. Benny Suprapto, Director of Secondary General Education; Theresia Pietersz, National Project Consultant; and Dr. Gordon Aylward, a consultant from Australia—were committed to the idea of a substantial onservice element. It has to be said that this was probably based less on the sort of theoretical considerations discussed earlier than on their combined experience of inservice teacher education projects in various parts of the world, where they had too often seen expensive but centrally based projects founder as soon as external funding terminated.[1]

An immediate question is: How can one possibly provide in-class coaching to many of thousands of teachers spread over such a large and often inaccessible area? What would be the cost and logistics involved in deploying the army of coaches required? PKG's solution was entirely appropriate to the situation: Draw coaches from the ranks of teachers themselves. Over the years there have been many modifications to the detail, but the basic model involves:

1. With the help of provincial and district authorities, use criteria including experience, qualifications, and tests to select a cadre of existing teachers from the secondary schools that were to be targeted.
2. Train this cadre of teachers at provincial, national, regional, and on occasion international workshops.
3. Return them to their own schools for a semester, where they practice for themselves the methods they have learned, and meet weekly in small groups to share experiences. National consultants make some inputs and monitor these meetings.
4. Accredit this cadre of trained and experienced teachers as "Trainers" or "Key Teachers" (*Guru Inti*). They now become responsible for running the main training programs in their provinces. These programs last one semester each and consist of
   a. A 2-week introductory workshop
   b. Meetings every Saturday throughout the semester
   c. Coaching visits by the trainers to teachers in their own schools during the week

There are also a mid-semester 1-week workshop and an end-of-semester 1-week workshop for reflection and transfer work.

5. Trainers and key teachers meet at annual national workshops to reflect on their coaching practice and to develop new content materials.

There are many more aspects to this vast professional development program than can be detailed here, but in summary what we have is a sort of two- or three-step cascade, but with a critical added element of feedback up the cascade, and continual monitoring by the national team of national and provincial workshops to maintain quality and guard against the classic dilution effects of standard cascade models.

Evaluation

Early evaluation of this program by Egglestone (1984) was based on classroom observations and assessments of student group[2] practical work in matched PKG and non-PKG classrooms, supplemented by collecting grades on nationally set examinations of content knowledge. They reported significant positive effects in the PKG classes on student active participation in the learning process, on their practical science problem-solving ability, and small but nonsignificant gains in the national science subject matter assessments. The last of these is important since teachers and administrators often expressed the fear that the extra time spent in PKG classes on practical work and constructive discussion might adversely affect the students' scores in the national tests. That students maintained expected levels of recall knowledge while experiencing a far richer program of constructivist teaching was an important finding.

Unfortunately this is not the simple happy end to the story. Recent diagnostic surveys (Blazely, Samnai, Rahayu, & Purwati, 1996; Sadtono, Handayani, & O'Reilly, 1996; Somerset, 1996) and an investigation assisted by Tony Somerset (Mahady, Wardani, Irianto, Somerset, & Nielson, 1996) have all indicated that those initial gains have not been maintained. Although none of these diagnostic surveys was able to correlate data on teachers' inservice experience with the classroom observations, the overall impression is given that the early effects of PKG on the quality of teaching have not been maintained. Mahady et al. (1996) suggest that possible reasons for this are:

1. The growth in size of the program, leading to a dilution of the influence of the central team
2. The loss from the program of the onservice visits to provide in-class support to teachers trying new methods
3. The addition of an extra step in the "cascade" from national level to classroom level

Of these, (1) may be the weakest hypothesis since PKG was already large, operating in 27 provinces, in 1984. Number (3) may be a factor, but it is one that is specifically addressed in the most recent form of the project where a two-way trickle-down, and feedback-up loop maintains the continuing development and monitoring of trainers at every level in the system. I suggest that it is (2), the loss of onservice visits, that must bear the main responsibility for a (temporary, we hope) loss of effectiveness of the PKG program. Here again we have evidence for the essential role that an onservice coaching element plays in effective change in teaching practice. An immense program, which was successful as long as the onservice work was maintained, appears to go into decline when that onservice is curtailed "due to cost." I will return to the issue of cost.

The Inservice-Onservice Model of CA in the United Kingdom

The other project to be considered is the professional development programme of the Cognitive Acceleration through Science Education (CASE) and Mathematics Education (CAME) projects in the United Kingdom. Cognitive Acceleration (CA) is an approach to teaching that focuses on the development of high-level (formal operational) thinking in young adolescents, with special activities set in science and mathematics contexts. These activities present students with cognitive conflict, and their effective delivery requires that the teacher generate the social construction of ways of thinking that have general value across many contexts. There is a strong emphasis on metacognition. The characteristics of teaching for CA are very different from the normal instructional pedagogy, whose main objective is the development of science concepts. Introducing CA into a school requires, in most cases, a radical reorientation of teaching philosophy, objectives, and methods and thus the professional development program is at least as important as the curriculum materials themselves. In particular teachers have to learn—against the trend imposed by a national curriculum consisting mostly of content objectives—not to worry about "covering" or "delivering" any particular science or math content during the Thinking Science and Thinking Mathematics lessons (Adey, Shayer & Yates, 1992; Adey, Shayer & Yates, 1995; Adhami, Johnson & Shayer, 1998) and to take much more time to manage extended discussions of difficulties encountered, justifications for positions taken, and opportunities to share insights amongst students.

CASE was initially run as a research experiment from 1980 to 1984. At this stage, there was small-scale professional development for about 12 teachers participating in the trials, and the design of this PD drew on the PKG experience and, later, on the emerging work of Joyce and coworkers described earlier. Thus onservice coaching was included from the start. The results of the original CASE work has been widely published (e.g., Adey &

Shayer, 1993, 1994; Shayer & Adey, 1993), but in summary it was found that, compared with matched control groups, students following the CA program (1) showed significantly faster cognitive development over the 2 years of the intervention; (2) 1 year after the intervention scored significantly higher on tests of science achievement; and (3) most remarkably, 3 years after the intervention scored significantly higher grades on nationally set examinations in science, mathematics, and English. This long-term far transfer effect of a cognitive intervention program made national news in 1991, just at a time when the political imperative in the United Kingdom to "raise educational standards" meant that schools were being publicly compared with one another on the basis of the national examination grades. The result was an immediate demand from schools for the materials and associated professional development to allow them to introduce Cognitive Acceleration.

Each year since September 1991 we have been starting a new cohort of schools on a 2-year CA PD program. Recently we have shifted the emphasis to working with local education authorities rather than with individual schools and by 2000 had been directly involved in PD programs for CA in between 400 and 500 schools. Since we insist on working with all of the science or math teachers in each school, the influence has been felt by many thousands of teachers.

In line with the main objective of this chapter, here I focus only on the onservice, coaching, element of the PD program. Each school buying into the 2-year program is entitled to five half-day visits by a CA tutor for coaching. Tutors fulfil the roles outlined previously, that is, they observe and offer suggestions for reflection and they help teachers to make explicit their implicit knowledge and so expose it as necessary to the process of conceptual change. Evidence of the effectiveness of this coaching process comes from a variety of sources:

1. Teachers themselves report the value of having a sympathetic observer guiding them as they change their practice.
2. CA tutors provide written reports back to the teachers they observe and track changes that occur between onservice visits.
3. In a long term follow-up study of schools that had participated in the 1994–1996 CA PD program, Adey, Bailey, Edwards, and Michael (1999) found that those schools that had not continued to maintain use of CA methods 2 years after the end of the induction were just those that had not taken up their quota of onservice coaching visits.

Conclusion

There are of course many differences between the two projects I have described. The sheer geographical scale of PKG dwarfs CASE; the pedagogical objectives of PKG are related to improving the quality of subject teaching, while those of CASE are more concerned with general cognitive development; and the funding available to CASE, through schools themselves, is relatively greater than that for PKG. But for the present purposes it is the common role played by school-based onservice work that is of interest. I believe that our experience of these projects over many years and the evidence available from them reinforces strongly the view derived from the theoretical bases described at the start of this paper and from the research evidence of Joyce et al., that in-school coaching is critical to the impact of staff development programs on student achievement.

Of course, the provision of onservice coaching in teachers' own classrooms is not without problems and difficulties. These include:

1. **Cost.**   Providing the kind of personal in-class attention that is characteristic of both of these large-scale projects is an expensive process. In the United Kingdom the CA tutors are paid as consultants, and despite efforts to make the system efficient by maximizing the number of lessons observed in one visit and where possible by visiting two schools in 1 day the CA PD program is expensive to schools. At least, with a high population density in England, travel times and distances rarely demand overnight stays. In Indonesia, in contrast, the tutors are regular teachers on (by Western standards) very low pay, but the time and cost involved in travel to remote schools and for getting tutors together for regular workshops can be considerable.

2. **Training the trainers.**   As the projects expand and the demands for onservice coaching increases, there is a growing pressure on the selection and training of tutors. Effective onservice coaches need to be experienced classroom teachers themselves, they need to have an unusually good understanding of the underlying theory and practice of the innovation they are introducing, and they need to be sympathetic but firm leaders. To find and train a large number of such paragons can present real difficulties. In (April 2000) the British Department for Education and Employment has charged us with the task of identifying between 12 and 20 trainers to introduce CA methods in 13 pilot Local Education Authorities in September 2000. This was feasible, but when the project expands nationally, our resources may be stretched.

3. **Coaching not inspection.**   In both the United Kingdom and Indonesia, teachers are very accustomed to the process of formal inspection, when the quality of their work is judged and reported upon. This is a stressful experience, and in some cases their prospects for promotion and their pay are dependent on the outcomes of such inspection. Thus when tutor-coaches first observe

them as part of a PD program, there is often nervousness and reluctance to experiment. It takes some time and sympathetic handling to demonstrate to teachers that the coaching process is very different from inspection and to create an atmosphere that actively encourages experimentation by the teacher.

None of these problems is insurmountable, but each needs to be considered in establishing a PD program that includes a significant onservice coaching element. The bottom line is: Do you want a professional development program that is cheap but ineffective? Obviously not. The only alternative, the only way to build a PD program that actually has a permanent and large-scale effect on students' learning, is to ensure that it includes coaching support for teachers in their own classrooms.

Notes

1. Information on the PKG project has been obtained directly by the author in his role as a minor consultant and reviewuator of the project from 1981 to 1984, and subsequently on many consultancy visits for the World Bank and Indonesian Ministry of Education. Opinions expressed are those of the author and do not necessarily reflect policies of those bodies.
2. In designing the revision, Jim Egglestone recognized that all practical work was conducted in groups of about six students and that assessing individual practical skills would be culturally inappropriate. His team thus developed practical problems to be solved by groups.

References

Adey, P., Bailey, M., Edwards, J., & Michael, N. (1999). *An archeology of a school improvement program—what is left after four years?* Paper given at American Educational Research Association, Montreal.

Adey, P., & Shayer, M. (1993). An exploration of long-term far-transfer effects following an extended intervention programme in the high school science curriculum. *Cognition and Instruction, 11*(1), 1–29.

Adey, P., & Shayer, M. (1994). *Really raising standards: Cognitive intervention and academic achievement.* London: Routledge.

Adey, P., Shayer, M., & Yates, C. (1992). Thinking Science - U.S. Edition. Philadelphia: Research for Better Schools.

Avgitidou, S. (1997). *Developing professional development in initial teacher training: a developmental model of student teachers' understanding of the relationship between theory and practice.* Paper given at EARLI, Athens.

Baird, J.R., Fensham, P.J., Gunstone, R.F., & White, R.T. (1991). The importance of reflection in improving science teaching and learning. *Journal of Research in Science Teaching, 28*(2), 163–182.

Blazely, L.D., Samnai, M., Rahayu, Y.S., & Purwati, R. (1996). *JSE science: Diagnostic survey* (10). Jakarta: Directorate of Secondary Education, Ministry of Education and Culture.

Brown, L., & Coles, A. (2000). Complex decision making in the classroom: The teacher as an intuitive practitioner. In T. Atkinson & G. Claxton (Eds.), *The intuitive practitioner*, 221–232. Buckingham: Open University Press.

Calderhead, J. (1993). The contribution of research on teachers' thinking to the professional development of teachers. In C. Day, J. Calderhead, & P. Denicolo (Eds.), *Research on teacher thinking*, 185–197. London: Falmer Press.

Cooper, C., & Boyd, J. (1999). Creating sustained professional growth through collaborative reflection. In C. M. Brody & N. Davidson (Eds.), *Professional development for cooperative learning* (pp. 49–62). Albany: State University of New York Press.

Egglestone, J. (1984). *An reviewuation of the PKG inservice-onservice teacher education project.* Jakarta: Directorate of Secondary General Education, Ministry of Education.

Guskey, T.R. (1986). Staff development and the process of teacher change. *Educational Researcher, 15*(5), 5–12.

Joyce, B., & Showers, B. (1980). Improving inservice training; the messages of research. *Educational Leadership, 37*(5), 379–385.

Joyce, B., & Showers, B. (1995). *Student achievement through staff development.* 2nd ed. New York: Longman.

Joyce, B., & Weil, M. (1986). *Models of teaching.* 3rd ed. Englewood Cliffs, NJ: Prentice-Hall.

Korthagen, F.A.J., & Kessels, J.P.A.M. (1999). Linking theory and practice: Changing the pedagogy of teacher education. *Educational researcher, 28*(4), 4–17.

Labaree, D.F. (1998). Educational researchers: Living with a lesser form of knowledge. *Educational Researcher, 27*(8), 4–12.

Mahady, R., Wardani, I.G.A.K., Irianto, B., Somerset, A., & Nielson, D. (1996). *Secondary education in Indonesia: Strengthening teacher competency and student learning* (1a). Jakarta: Directorate of General Secondary Education, Department of Education and Culture.

McMahon, A. (2000). The development of professional intuition. In T. Atkinson & G. Claxton (Eds.), *The intuitive practitioner* Buckingham: Open University Press.

Mevarech, Z.R. (1995). Teachers' paths on the way to and from the professional development forum. In T.R. Guskey & M. Huberman (Eds.), *Professional development in education: New paradigms and practices*, 114–132. New York: Teachers College Press.

Raw, A. (2000). CASE Proven. *Teaching Thinking, 1*, 47–50.

Sadtono, E., Handayani, & O'Reilly, M. (1996). *English diagnostic survey with recommendations for inservice training program for SLTP teachers* (8a). Jakarta: Directorate of Secondary Education, Ministry of Education and Culture.

Schon, D.A. (1987). *Educating the reflective practitioner*. San Francisco: Jossey-Bass.

Shayer, M., & Adey, P. (1993). Accelerating the development of formal operational thinking in high school pupils, IV: Three years on after a two-year intervention. *Journal of Research in Science Teaching, 30*(4), 351–366.

Somerset, A. (1996). Juniow Secondary School Mathematics: Diagnostic Survey of Basic Number Skills. Jakarta: Directorate of General Secondary Education, Ministry of Education and Culture.

Tomlinson, P. (1998). *Implicit learning and teacher preparation: Potential implications of recent theory and research*. Paper given at British Psychological Society Annual Conference, Brighton.

Wilson, S.M., & Berne, J. (1999). Teacher learning and the acquisition of professional knowledge: An examination of research on contemporary professional development. In A. Iran-Nejad & P.D. Pearson (Eds.), *Review of research in education, Vol. 24* 249–306. Washington, DC: American Educational Research Association.

Chapter 11

Managing Teacher Development
The Changing Role of the Head of Department in England

JUSTIN DILLON

Teacher development is not something to be done solely in private behind closed doors, at least not if it is to be effective in changing teachers' practice (Joyce & Showers, 1988). The skills and knowledge required for effective teaching develop and grow in a range of settings and are constrained by a wide variety of factors ranging from personal circumstances to systemwide features such as the curriculum or assessment policies. In England, which has undergone substantial education reform, particularly since the mid-1980s, there have been many attempts to provide professional development opportunities for individuals, for groups of teachers, and for the whole teaching cohort. One feature that characterizes the climate in which teachers work is the focus on standards and outcomes rather than on processes. A government-funded body, the Teacher Training Agency, has produced a normative set of standards that are meant to guide teachers and professional development "deliverers." These standards are already the subject of interest of a range of countries around the world.

Since the mid-1980s, researchers have investigated the success of specific projects or courses and have suggested lessons for a wider audience, while others have looked at the general circumstances within which teachers operate. What seems clear is that teacher development *is* possible but is likely to be hampered by a range of barriers to change, many of which show no sign of disappearing. In this chapter I focus on two pieces of research with which I and colleagues have been involved that examine the current state of professional development in England at a systemwide and at a more school-focused level. One study involves a nationwide survey of science teachers' needs and wants, whereas the other is an in-depth study of the management of teacher development. I argue, using evidence from these and other studies, that normative models of teacher development are totally unrealistic and inappropriate for addressing the needs of teachers and schools.

Background

In England, education is compulsory for children ages 5–16. Most children attend primary schools (of which there are about 26,000 in the country) until they are 11 and then transfer to secondary schools (of which there are about 4,000). In the former, they are likely to be taught by the same teacher for all subjects for 1 year before moving onto another teacher in September. In secondary schools, children have different teachers for different subjects and could well have the same teacher at several times in their schooling. This chapter refers only to those teachers in state schools, those funded by the government, which look after around 93 percent of the country's children.

For various reasons, a national curriculum, made up of a range of subjects studied each year, was introduced in 1988. National assessment, which takes place at the ages of 7, 11, 14, and 16, is used to monitor attainment as well as to compare schools. Science is one of three core subjects that students must take throughout their years of compulsory education. Partly as a result of this "core subject" label, secondary science teachers, who are usually graduate scientists with an education qualification, have a relatively high status compared with their art, history, geography, and modern language colleagues.

Newly qualified secondary school teachers would expect to teach about 36 out of 40 periods (lessons) each week. A science teacher would normally be part of either a science department or a single-subject (biology, chemistry, or physics) department led by a Head of Department (HoD) (Departmental Chair in the United States). The HoD, as well as being paid for his or her responsibility, would expect to be teaching less than an ordinary classroom teacher—say, 32 periods each week. It has been argued that the pyramidal hierarchy of the English education system owes much to the influence of military service on a generation of young adults during and after World War II.

The surface features of the current role and responsibility of the HoD do not appear to have changed greatly over the years. However, a growing climate of accountability and managerialism in English schools has led to fundamental changes in the day-to-day existence and experience of many HoDs. In particular, the role of the HoD in teacher development, through monitoring, appraising, and setting targets, is now radically different. As an example of the climate of expectation that surrounds HoDs, one need look no further than the standards set by the Teacher Training Agency (TTA), a government funded body for subject leaders (the name given to heads of department in secondary schools and subject coordinators in primary schools). The standards define "key outcomes" of the job, which include "teachers who: work well together as a team; support the aims of the subject . . . [and] are dedicated to improving standards of teaching and learning" and "the ability to lead and manage people to work as individuals and as a team towards a com-

mon goal" (Teacher Training Agency, 1997, pp. 4, 6). The utility of normative models of management, which fail to problematize the roles and responsibilities of teacher leaders, is highly questionable. The imprecision and
naïveté of such models are neither enabling nor informative. The reality of
HoDs—the changing role and the consequent impact on the management of
teacher development—are the foci of this chapter.

The Issues Affecting Science Teachers' Professional Development

Heads of Department can only work within the constraints of the system. The
current status in England with respect to science teachers' experience and
expectations toward professional development can be gauged from a study
commissioned by the Office of Science and Technology in 1999 (Dillon,
Osborne, Fairbrother, & Kurina, 2000). The study's findings are based on
research with 20 focus groups involving more than 150 teachers from 50
schools in five regions of England and a questionnaire survey of a randomly
selected sample of 1,973 primary and 735 secondary state schools in England. The study looked at a range of key issues: teachers' qualifications in
science, their opinions of their preservice training, their current teaching profile, their experience of inservice training during the year, their opinion of
that training and advice, and their desires for improvements in the quality of
what professional development was offered to them.

In England, most secondary school teachers take a 1-year Postgraduate
Certificate in Education (PGCE) at a university department of education.
English science teachers, in line with those in some U.S. studies (see, e.g.,
Luft & Cox, 1998), feel well prepared for teaching some or all of the science
curriculum at the end of their preservice courses. However, they are very crit-
ical of the support that they receive during their careers.

A particularly worrying feature for Heads of Department was that the
teachers in the study were not engaged in a subject-related, classroom-based,
systematic process of Continuous Professional Development (CPD) matched
to their individual needs. Interestingly and critically, "many teachers in the
focus groups did not fully appreciate the term CPD." Although in the 1980s
teachers had 5 days of holiday removed and replaced with 5 days devoted to
training, there was significant widespread dissatisfaction with the use made
by schools of those days. Teachers felt that there was too much focus on
administration or on whole school issues, such as literacy and numeracy, and
too little focus on personal, professional development.

Another disturbing feature for Heads of Department was that teachers
were critical of the existing "appraisal arrangements for identifying their
individual strengths, and of just how little say they had in their individual
CPD, or the courses that they did attend." Although most teachers had
received some form of inservice education and training (INSET) during the

preceding school year, it was mainly from colleagues in their own school rather than from other sources of inservice education and training. In their view, INSET at present was insufficiently focused on their individual needs, involved too little in the way of practical activity rather than pure theory, and too rarely permitted interaction with other teachers.

What the teachers *did* want were more opportunities, "not only to share experience and good practice with colleagues in their own school and in other schools, but also to compare their practice with others in order to identify their individual CPD needs." Specifically, they wanted higher quality INSET and more emphasis on classroom-focused support. In terms of topics, the teachers prioritized subject knowledge, pedagogy, pupil learning, and classroom management. Dillon et al., 2000 p 21.

Around half of the teachers in the study had made use of local government advisors (although 31% of those that had received advice rated it as "poor") and only 25 percent had been to their local teachers' center (of those, 42% described the help as "poor"). Of all our data, these negative findings were perhaps the most surprising.

The main professional association, the Association for Science Education (ASE), had provided advice for 56 percent of the teachers and was very positively reviewed. Teachers subscribe to the ASE and so might be more likely to review it positively than they would if it were a free service. The role of professional organizations is somewhat underresearched in England despite their importance in influencing both government policy and teacher practice.

One of the major issues in science education in England has been the effect of the move from separate science teaching (biology, chemistry, and physics) to "balanced science." Many teachers do not have an adequate background in all three science subjects. Among those teachers teaching science topics to students ages 14–16, 26 percent of those teaching biology topics did not have an A-level (taken at age 18) in the subject, nor did 13 percent of those teaching chemistry topics and 29 percent of those teaching physics topics. Furthermore, 39 percent of those teaching biology topics did not have a degree in biology, 51 percent of those teaching chemistry topics did not have degree in chemistry, and 66 percent of those teaching physics topics did not have a degree in physics. Not surprisingly, teacher confidence, one of the major factors affecting teaching quality, varied from topic to topic.

Another area in which successive governments have tried to innovate in science education has been in the use of computers. The Department for Education and Employment has spent significant sums of money equipping schools with computers and on setting up a National Grid for Learning (NGfL). Despite this investment, the level of use was reportedly rather low. From the questionnaires returned, the NGfL was used only "rarely" by 72 percent of teachers. Whereas 42 percent of primary teachers reported using computers "often" in their science teaching, only 9 percent of secondary science teachers could say the same.

The amount of money available for INSET varied widely from school to school. The mean total amount available in the secondary schools was £14,730. The amount of money available per member of staff ranged from £50 per teacher to £1500. The average amount was £304. The head (principal) or a senior colleague generally makes decisions about how INSET funds are allocated. Most heads often determine INSET needs in line with School Development Plans with individual requests next in line as a determining factor.

Head teachers differentiated between the needs of less and more experienced teachers. For inexperienced teachers of science, the head teachers rated Teaching Skills, How Students Learn, Raising Achievement, and Class Management as the most important. For more experienced teachers head teachers viewed Raising Achievement as by far the most important topic, with How Students Learn and Middle Management some way behind. The major constraints on the provision of adequate professional development, according to head teachers, were, unsurprisingly, a lack of time and a lack of finances.

We concluded that there was an urgent need to examine carefully the effectiveness of the current provision for *personal professional development* of teachers of science. In particular, more time and funding need to be allocated to personal, professional training rather than institutional imperatives. In England, there needs to be a major cultural shift within the profession to make subject-specific, classroom-based continuous professional development the norm. At present, for a significant number of newly qualified teachers of science, if not the great majority, their individual needs for subject-based CPD take second place to whole school issues, and any support they do receive diminishes rapidly after the first year of their careers.

In concluding, we argued that school managers have a responsibility for the creation and maintenance of a pro-CPD culture in a school. Management should have the capability to sustain and nurture the subject-related expertise of its teaching staff effectively within the existing very real constraints on science teachers' CPD that the study has highlighted, namely, the provision of time and funding, an appropriate reduction in workload and fatigue, and the necessary supply cover for teachers to attend external courses.

Government Initiatives

One strategy adopted by previous governments and reinforced by the current government is mandatory appraisal for teachers. The philosophy behind the scheme is that experienced teachers are able to identify the effectiveness and the competence of their colleagues and can suggest the most appropriate professional development for them. In practice, heads of department observe a teacher and then talk through the teacher's strengths and weaknesses, setting targets for improvement and suggesting strategies and professional development needs. In practice, appraisal has been a major failure (OfSTED, 1996). Specifically:

- The impact on teaching standards and learning has not been substantial.
- In a majority of schools appraisal has been too isolated from planning for inservice training (INSET) and for school development.
- The classroom observation of teachers has been variable in quality, and this has led to appraisal interviews being insufficiently focused on the improvement of teaching.
- Schools have chosen not to link appraisal to pay and promotion, even indirectly.

The emphasis in appraisal has too often been the institution and not the individual. A second issue has been the inappropriate and inadequate nature of staff development available for teachers. The current government, picking up on the last point listed by OfSTED, has deliberately linked appraisal with performance. Performance-related pay is likely to be one of the major controversies in education in England in the first decade of this century.

The Impact of Educational Reform on Teachers and on Their Professional Development

For several years I have been studying the impact of government reform on heads of department. The impact of the National Curriculum has been profound (see, e.g., Donnelly & Jenkins, 1999). For example, the pressure of an overcrowded curriculum has caused teachers to adopt a transmissive mode of teaching (Hacker & Rowe, 1997; Osborne & Collins, 1999).

Helsby and Knight (1997) identified three postreform changes that are relevant to the area of professional development: "an obvious and pressing need for new learning, to build or extend knowledge of subject matter; to develop pedagogical processes appropriate to the new requirements; and to foster mastery of teaching to, and assessing against, the AT [attainment target] statements" (p. 146). However, they go on to point out, "The changes in the formal structures of in-service education and support for teachers (INSET) which have accompanied the educational "reforms" of recent years, have seriously restricted the opportunities for personal, professional development" (p. 149). According to Helsby and Knight, INSET is now "heavily managed from the center within tight budgetary constraints" (p. 149). The onus for staff development now lies more with schools than at any time in the last 30 years.

The implementation of the National Curriculum coincided with responsibility for managing financial resources being devolved to schools. This resulted in a lack of any systemwide priorities at local or national level. For instance, Her Majesty's Inspectorate (HMI) (1992) commented, "Teachers attended a range of courses but with many schools' receiving devolved INSET funding, much of the INSET has been school-based. . . . Overall, however, the systematic identification and prioritization of INSET needs, both individual and departmental, was not sufficiently common" (p. 27).

There is also further evidence that schools and Local Education Authorities have not invested enough in training for middle management, particularly those aspects of the job that focus on staff development, appraisal, and teamwork. HMI (1992) concluded, "Schools and LEAs would be wise to invest a greater proportion of INSET provision in the training and support of Heads of Science" (p. 31).

The Reality of Teacher Development

I have previously outlined a range of factors that might encourage successful teacher development, including:

- Initial disturbance or dissatisfaction
- Time to reflect on existing strategies
- An evidence base of successful teaching strategies based on models of learning
- Feedback on classroom performance
- Encouragement from managers
- A feeling of personal growth
- A sense of ownership of innovation (Dillon, 2000)

Simply listing supposedly desirable factors is nothing more than an academic exercise unless there are mechanisms in schools for the factors to be facilitated. Without informed, committed, and skilled managers, the prospects that teachers will be able to engage in systematic, continuous, personal professional development are almost nil.

In a study conducted with Heads of Department in nine schools in a large city in Southeast England, I have found evidence that throws doubt on the likelihood that the culture of CPD is likely to change radically in the near future (Dillon, in preparation). In the light of the findings from that study, I will address some of the issues that affect the factors just listed.

Initial Disturbance or Dissatisfaction

Bell and Gilbert (1996), in their study of New Zealand science teachers, found that one of the critical stages in teacher development was the recognition that an aspect of personal practice was problematic. This echoes Nancy

Davis' (1996) finding that a sense of dissatisfaction can lead to a desire to change an aspect of a teacher's practice.

However, in the current climate of accountability and managerialism, which pervades schools in England, it is unlikely that teachers will be eager to admit too readily to problems with their practice. Performance-related pay, introduced in September 2000, means that teachers are set targets on an annual basis. Admitting to failure could well mean that a teacher does not receive a pay rise. This is hardly the climate to foster open admission of problems.

As I have pointed out elsewhere (Dillon) "in departments which have experienced staff, successful students and an established social structure, it is very difficult to create a sense of dissatisfaction with existing practice" (p. 94–109).

Time to Reflect on Existing Strategies

The overwhelming pressure teachers and their managers are faced with is that caused by lack of time. The busyness that characterizes schools in England is the result of almost constant change—curriculum, assessment, funding, to name but a few. Managers spend more time on administration than they do on person management. Few involved in education have the opportunities for reflection, which, as Dewey (1933) says, involves "a state of doubt, hesitation, perplexity, mental difficulty, in which thinking originates, and . . . an act of searching, hunting, inquiring, to find material that will resolve the doubt, settle and dispose of the perplexity" (p. 12).

One of the participants in my study, an experienced Head of Department, summed up what had happened to her before I interviewed her:

> Oh God, today has been a really bad day. I've been in tears today. That hasn't happened for a long time. It's been horrible. I've just felt that all day today I have been like . . . I get strings of children sent to me consistently for behaviour [. . .] I'd been on the go since eight this morning, . . . , consistently following up cases of bad behaviour, right, I taught two lessons this morning and then I observed Gerard after break. And that wasn't too bad and then it started really badly about 12:00 P.M. and I spent the whole of my own lunchtime dealing with this awful class and all they were saying was what rights they had. Ah, you know, the street talk and none of them owned up to this and it made me feel sick

Lack of time to find out about new ideas limits the discourse of middle managers to repeating the rhetoric of their own managers: targets, performance, appraisal, and improvement. Managers' abilities to describe what is required in concrete terms at anything more than a general level ("students on task," "safe environment") can lead to frustration on the part of teachers who are vague about what is required and on the part of managers who feel help-less to provide real help.

An Evidence Base of Successful Strategies Based on Models of Learning

Evidence that teachers access educational research is hard to find, to say the least. This is not because researchers have ignored teachers as consumers of their findings (see, e.g., Adey & Shayer, 1994; Driver, Squires, Rushworth, & Wood-Robinson, 1994; Monk & Osborne, 2000; Novak & Gowin, 1984). In general, "teachers are forced to rely on local networks of informal contacts, either in-school or between schools" (Osborne & Dillon, 1999, p. 3).

It should be added that although a growing body of research has taken place into school effectiveness and school improvement, the validity of the findings has been questioned. There is a perception that there is an education management market that has so far failed to deliver a quality product (Gunter, 1997).

Feedback on Classroom Performance

Joyce and Showers' (1988) meta-analysis of the effectiveness of inservice education showed that success depends on *long-term*, classroom-focused coaching involving feedback. However, I found little evidence that teachers watch each other frequently or systematically, and when they do it is often part of a formal, short-term, appraisal system. As one HoD, Anne, put it:

> The plan . . . was that each member of staff was observed twice by two sep-arate people. . . . I'd be looking at doing it termly. And really, if everybody was involved then it's a case of doing one and maybe being looked at once. I started off by doing "buddy pairs" . . . so that you saw and watched some-body else so that the two of you can work together. Then you can change the "buddy pairs" . . . or you might say, I can't work out how to do Science 1,[1] the investigative work, you might match up with somebody then who'd be a bit stronger.

Appraisal of teachers by managers, of which the above example is one model, became statutory in England in the 1990s, although it has not had the impact that was originally hoped. Formal appraisal usually involves focused observation and other data gathering followed by discussion. In reality, because HoDs teach for the majority of the week, the opportunities for appraisal are limited. The idea of mutual observation is also becoming more common, although the time constraints for teachers without significant responsibility are even greater than for HoDs. Mike, one of the HoDs in my study, put it:

> I try to observe everybody at least once a year formally. There's the book monitoring: You can see what the sort of standard of the students' work is. There's the homework setting and the defaults from that. And I think you can pick up more when people put in no defaulters than when they put in a whole list. And so it's a question of saying, well didn't you get any defaulters last week? And then, if they say no, you say, well what did you set, you know, have you collected it in, got it marked. Why are you getting no defaulters and I'm getting about half-a-dozen?

For various reasons, many teachers have deployed a range of strategies to "block, deny or distort feedback from the rare observations that they receive from inspectors or other colleagues" (Dillon, 94–109). As I have pointed out earlier, "[T]he changes in the curriculum and the assessment policies in England, which have failed to shake the reliance on memorization as a learning strategy, have failed to create much dissatisfaction in teachers' own practice. Teachers are in danger of becoming resistant to criticism as a strategy for maintaining professional pride" (Dillon, 97).

During one of my interviews, an experienced HoD, Alan, was questioned as to how his school was preparing for a forthcoming inspection: "Well Senior Management say they're going to get all the "top bods"[2] in to tell us what to do but I think some of us, and I think myself included, are saying 'well you know, we'll do what we can but we're going to carry on doing our proper job and not let OfSTED get in the way too much.' " To me, the idea that experienced teachers and managers are simply told what to do is an example of the "discursive working of the new managerialism" (Reay, 1998, p. 188). The mechanistic, technocratic approach to schools and to education reform has set an agenda for change that is disempowering and disenchanting.

Encouragement from Managers

The importance of support for teacher development from senior management has been recognized for some time (see, for example, Adey, Dillon, & Simon, 1995). Getting the balance right between support and pressure is a key aspect of teacher development. In England, however, it does seem that the role of the HoD has shifted more toward pressuring rather than supporting. HoDs are now "proxy-managers"—carrying out school policy (and government policy) at a distance—using appraisal as a means of quality assurance (Dillon, 1997). As one of the HoDs indicated, the period of adjustment for this process has been minimal, the training variable, and the penalties for failure high, particularly for the less powerful groups in schools: "[We] are more accountable than [we] ever were before. Changes are taking place . . . overnight and we're having to . . . implement them more-or-less as we speak and there are so many changes that people are threatened by that . . . especially the young teachers.

A Feeling of Personal Growth

The aspects of teaching that appealed to me throughout my classroom career was what Huberman (1989) recognised as the opportunity for "tinkering" with resources and teaching styles. In Huberman's study, teachers that engaged in tinkering were likely to be "more satisfied" later in their career than those who did not adapt. Huberman also identified other factors that were predictors of "satisfaction"—the first being the ability to "change one's role when one begins to feel stale" and the other comes from a realisation that one's students are actually learning something. In the light of the issues raised earlier, opportunities for personal growth appear to be limited for many teachers. The evidence from my research is that teachers, particularly in small schools, are faced with increasing isolation and pressure, which are more likely to result in feelings of inadequacy rather than growth.

A Sense of Ownership of Innovation

One of the taken-for-granteds about teacher development has been the feeling that teachers needed to feel committed to change (to "own" it) for it to be implemented, although some research has questioned that necessity (Adey, Dillon, & Simon, 1995). Personal control over the change process should be a desirable situation if only on democratic and equity grounds rather than simply psychological grounds. My understanding is that teachers use a range of micropolitical processes to exert ownership on change, whether it is through interpretation of the curriculum or through deliberate refusal to participate in activities.

Gold and Evans (1998) express the view that "excessive micropolitical activity within a school may be indicative of blocked or ineffective decision-making routes" (p. 22). They go on to say that "whatever the cause of the excessive micropolitical activity, those with management responsibilities within a school need to be aware when they are overactive and to make some basic decisions about whether to use or ignore the unofficial structures." However, it must be borne in mind that: "Decision-making is not an abstract rational process which can be plotted on an organizational chart; it is a political process, it is the stuff of micro-political activity" (Ball, 1987, p. 26). This position does not seem to be recognized by those giving advice to HoDs (see, e.g., Gold, 1998). Normative models of teacher education are more likely to be subverted if teachers perceive themselves as being molded into particular forms.

Gender

The issue of gender in management, which was not one of the factors listed earlier, has been studied but is still little understood in terms of teacher development and its management in England. As more women become managers and more men become managed, there is a continuing need to appreciate the gender issues involved. My feeling is that research into gender issues needs to focus more than it has upon providing rich descriptions and critical analysis of the situations in which women find themselves as managers. Stereotypical views of women as managers are still too common. One of my participants articulated a view that is rarely found in the literature:

> I'm very careful how I deal with the men 'cos I think [. . .] and I'm not saying this in front of you in any way to insult male populations, but I feel that with men you've got to be careful. They're not used to women managing them a lot of the time and um . . . and I don't want to ruin their egos. I don't want to denigrate them and make them feel small. But sometimes they see me with the kids and how strong and tough I am with them. Not necessarily succeeding, but I am tough with them and I suppose maybe feel a little bit threatened by that. . . . So I do, I do deal with the men differently, the woman I tend to have a great relationship with. They just accept everything. I have no problems [. . .]

Another female HoD, Kelly, spoke of the differences between men and women.

> I think women tend to be more flexible and more adaptable in terms of what they're prepared to accept and do sometimes see alternative viewpoints though there are obviously still some women who see something and decide against it also. But I mean there are big disadvantages having a woman because quite a few of them have got young children and so they have the commitment to childcare that men don't always have.

Final Comments

The focus of this chapter has been on the challenges faced by today's Heads of Department. However, I think that a range of issues found in England have a more widespread relevance. From my experience of education systems around the world, there are many similarities in the way professional development is conducted in global communities:

- The tendency to ignore contextual constraints on teacher change
- The indifference to teachers' needs and wants and the simplistic assessment of the effectiveness of teacher development
- The disempowering imposition of standards and procedures

What I have tried to do here is to use the English context as an exemplar rather than to say that England is unique in all aspects of teacher development. However, it is unique in some aspects. The suddenness of change from extreme hands-off to extreme hands-on in terms of the government's attitude toward the education system has rarely been witnessed elsewhere. The consequent grinding of gears has fragmented the teaching profession in a way that has not happened in many, if any, other countries. These contextual factors have led to a range of micropolitical realities that contrast with the mythology of professional development legislature and guidance.

The culture of accountability and managerialism has major implications for equity issues. The use of power and control to effect changes in the education system relies on traditional attitudes and structures coupled with the reviewing of individual agency as a means of (self)-improvement. The model of professional development that emerges from a study of English teachers is one of spasmodic, unplanned, and poorly funded support set in a virtual framework of standards and outcomes.

The success of the professional development program associated with CASE (Cognitive Acceleration through Science Education) (see Philip Adey's chapter in this volume) provides some lessons. CASE is based on a sound theoretical foundation, a clearly articulated and replicable teaching approach, trialled and tested support material, classroom-focused coaching, and feedback over a 2-year period. It is not cheap but it appears to work.

The extent to which models of professional development are transferable is clearly an issue in a book such as this. My own feeling is that although there are many differences between teachers and education systems from state to state and from country to country, some of the challenges are generic.

What can other countries learn from this work? The issue that should trouble researchers and professional developers alike is the seemingly constant reinvention of the wheel that seems to take place all too frequently. However, another message is clear—models of teacher development that are based on simplistic, normative, technicist ideas are unlikely to be effective. Teacher development needs support from managers, but if teaching is complex then managing teachers is even more so. Teachers and their managers need to see development as multifaceted, progressive, and dependent on individual personalities, psychology, and politics—both macro and micro. These lessons apply in education systems where there are few middle managers as well as in those, like England, where there are many.

Management is partly about confidence as well as competence. The inadequate training that Heads of Department receive, the lack of time that they have to manage, the ethical dilemmas facing them, the pressure on from parents, heads, inspectors, and colleagues all conspire to make the job difficult, unsatisfactory, and impossible to finish. The job of the Head of Science is possibly the most demanding of all middle managers, and there is little sign that that will change in the future.

Notes

1. This refers to a particular section of the National Curriculum followed in Anne's school.
2. That is, staff from the local education authority who have a role in monitoring standards in schools.

References

Adey, P.S., Dillon, J.S., & Simon, S.A. (1995). *School management and the Effect of INSET*. Paper presented at the European Conference on Educational Research, Bath.

Adey, P., & Shayer, M. (1994). *Really raising standards*. London: Routledge.

Ball, S.J. (1987). *The micro-politics of the school*. London: Routledge.

Bell, B., & Gilbert, J. (1996). *Teacher development: A model from science education*. London: Falmer Press.

Davis, N.T. (1996). Looking in the mirror: Teachers' use of autobiography and action research to improve practice. *Research in Science Education, 26*(1), 23–32.

Dewey, J. (1933). *How we think: A restatement of the relation of reflective thinking in the educative process*. Chicago: Henry Regnery.

Dillon, J. (1997). *Managing teacher development: The role of the Head of Department in England*. Paper presented at the European Association for Research in Science Education Summerschool, Barcelona.

Dillon, J. (in press).—Managing science teachers development. In R. Millar, J. Leach, & J. Osborne (Eds.), *Improving science education*: Buckingham: Open University Press. pp 94–109

Dillon, J. (in preparation). The Role of Middle Management in Science Teacher Development in Schools. Unpublished PhD thesis, King's College London.

Dillon, J., Osborne, J., Fairbrother, B., & Kurina, L. (2000). *A study into the professional views and needs of science teachers in primary and secondary schools in England*. London: King's College London.

Donnelly, J. F., & Jenkins, E. W. (1999). *Science teaching in secondary school under the National Curriculum*. Leeds: Centre for Studies in Science and Mathematics Education, University of Leeds.

Driver, R., Squires, A., Rushworth, P., & Wood-Robinson, V. (1994). *Making sense of secondary science*. London: Routledge.

Gold, A. (1998). *Head of Department: Principles in practice.* London: Cassell.

Gold, A., & Evans, J. (1998). *Reflecting on school management.* London: Falmer Press.

Gunter, H. (1997). *Rethinking education: The consequences of jurassic management.* London: Cassell.

Hacker, R.J., & Rowe, M.J. (1997). The impact of National Curriculum development on teaching and learning behaviors. *International Journal of Science Education, 19*(9), 997–1004.

Helsby and Knight (1997) Continuing Professional Development and the National Curriculum in Helsby, G. and McCulloch, G [eds], Teachers and the National Curriculum, London: Cassell, pp. 145–162.

Her Majesty's Inspectorate (HMI). (1992). *Science: Key stages 1, 2 and 3. A report by H M Inspectorate on the Second Year, 1990–91.* London: HMSO.

Huberman, M. (1989). *Teacher development and instructional mastery.* Unpublished paper presented at the International Conference on Teacher Development: Policies, Practices and Research, Toronto: Ontario Institute for Studies in Education.

Joyce, B., & Showers, B. (1988). *Student achievement through staff development.* New York: Longman.

Luft, J.A., & Cox, W. (1998). *Final report: A report on preservice and mentoring programs in Arizona for mathematics and science teachers.* Arizona Board of Regents: Eisenhower Mathematics and Science Program.

Monk, M., & Osborne, J.F. (Eds.). (2000). *Good practice in science teaching: What research has to say.* Buckingham: Open University Press.

Novak, J.D., & Gowin, D.B. (1984). *Learning how to learn.* Cambridge: Cambridge University Press.

Office for Standards in Education (OfSTED). (1996). *The appraisal of teachers 1991–1996.* OfSTED reference HMR/18/96/NS. London: Ofsted.

Osborne, J., & Collins, S. (1999). *Pupils' and parents' views of the role and value of the science curriculum.* Paper presented at the British Educational Research Association Conference, Brighton, September.

Reay, D., (1998). Micro-politics in the 1990's: staff relationships in secondary schooling, Journal of Education Policy, 13 (2) pp 179–196.

Teacher Training Agency (TTA). (1997). *National standards for subject leaders—annex.* London: TTA.

Chapter 12

Key Issues and Contextual Factors in Professional Development of Preservice Science Teachers

Perspectives from Israel, Greece, Italy, and the United States

URI ZOLLER AND DAVID BEN-CHAIM

In view of the overly high expectations of people in a world of conflicting/ competing values and finite unevenly distributed resources, modern life has turned into a continuous process of problem solving and decision making, or decision selecting, from either available or as-yet-to-be generated options (Zoller, 1991). However, although science and technology may be useful in establishing what we *can* do, neither of them (solely or jointly) can tell us what we *should* do. The latter requires the application of value of judgments by socially responsible, rational citizens as an integral part of their critical system thinking capacity, which, in turn, requires *evaluative thinking* by capable, rational science-literate citizens (Zoller, 1990, 1993, 1996). Thus, a major goal of contemporary science education is the development of the students' higher-order cognitive skills (HOCS) of reasoning, critical thinking, problem-solving, and decision-making capabilities in the context of both the specific content and processes of science and the reality-based science-technology-environment-society (STES) interfaces, so they can be effectively functioning citizens (Zoller, 1990). The ultimate goal is autonomous, science-literate and educated learners who are capable not only of *knowing* but also of *thinking*, meaning the development of students' HOCS capacity (Zoller, 1993, 1997).

Striving for this goal means developing skills that would enable students to participate effectively in the decision-making, problem-solving process of our democratic complex science- and technology-based society. It also means a shift from "covering material," via algorithmic teaching, for the sake of knowledge acquisition to enhancing the critical thinking of learners via HOCS-oriented teaching, for the sake of rational decision making and responsible action. Such a paradigm shift means a different conceptualization and, eventually, goals—both reflected in the teaching-learning process—which, in turn, requires different models of science teacher education with consequential results, effects, and impact on the professional development of the prospective teachers as well as their professors involved in these preservice (and later inservice) teacher training programs.

Reform advocates envision such a targeted dramatic change in science education occurring via a longitudinal systemic effort within which *all* components of the educational arena are involved and work together toward a common goal. This change is at the heart of the current educational reform movements (National Research Council, 1996; Zoller, 2000).

While strengthening our students' higher-order thinking skills is the primary goal of current reform efforts in science education, science teachers and science teaching are the means by which we attain this goal. This objective constitutes an unprecedented challenge to the traditional conceptualization of the science teaching enterprise, where the major goal was to transmit information about the discipline. Consequently, implementing reform means a change in policies and programs and applying new teaching strategies and assessment methodologies that are consistent with the Standards (Bybee, Ferrini-Mundy, & Loucks-Horsley, 1997; Zoller, 1993, 1996, 1997, 2000). This, in turn, requires corresponding new preservice science teacher programs (Simmons et al., 2000) and, in accord, different emphasis concerning the teachers' professional developments.

Students and teachers should not only actively participate in the teaching-learning process, but should also become partners in this process if science education reform is to succeed. For teachers of science the shift envisioned by the reform means:

- A change of focus from teaching/instruction to learning; that is, although both instructional design and how we teach are important, the more pressing issue is what is actually learned by the students
- Application of an integrated approach in dealing with real-life problems. This format requires that students conceptualize fundamental and/or interdisciplinary science topics (e.g., dynamic equilibrium, reversible and irreversible process, exponential growth) and sharpen their HOCS (Zoller, 1997, 2000).

The teaching of facts and imparting of disciplinary/compartmentalized knowledge results only in lower-order cognitive skills (LOCS) learning (NRC, 1996). In short, the essence of the current reform is a shift from the LOCS *teaching* to HOCS *learning*, from algorithmic teaching to evaluative thinking in the STES context (Zoller, 1993, 1997).

The advocated LOCS-to-HOCS switch constitutes an alternative to the existing (traditional) model of teaching and learning (Bybee et al., 1997;

Zoller, 1993, 1996, 1997). The main features of the new HOCS-oriented science teaching "model" are:

- A holistic, systemic, *interdisciplinary* approach as the guiding construct.
- Independent, active inquiry-based learning rather than passive knowledge acquisition where the learner/researcher constitutes an integral part of the investigated system.
- A personal involvement and responsible action on the part of the learner (Zoller, 1993).
- The educational objectives and students' (not disciplines') needs should be the main determinants of the teaching and assessment strategies of sciences and STES-oriented courses.
- The essence of the learning process is relevant question asking/problem raising and investigation by the student learner to develop a position, make a decision, and take action accordingly.

This model is guided by the ideal of the educated person: one who has (1) the ability to be engaged in HOCS-based activity both in the study of science and in dealing with everyday life problems, (2) the knowledge basis relevant to these problems, (3) the ability to select and effectively use *the relevant information*, and (4) the motivation and self-confidence to act accordingly and take responsibility. In short, an educated person is able to think rationally, logically, reflectively, and consequentially before deciding what to accept (or reject), determining what to do (or not to do), and taking the proper action (Zoller, 1997).

Science education plays a key role in preparing students for informed, intelligent, and responsible participation in the STES-related decisions of our democratic society. Therefore, HOCS such as question asking or generating, problem solving, decision making, and critical system thinking—all of which require *evaluative thinking*—should become legitimate important learning outcomes to which good science teachers should aim (Zoller, 1995). These skills constitute the core of the assembly of performance processes needed for coping with previously unprecedented complex problem situation and conceptual understanding transfer and system (inclusive) thinking, with respect to both the science disciplines and real-life problems within STES and their interrelationships context (Zoller, 2000).

Accordingly, the thesis of this longitudinal research-based chapter can be summarized as follows:

1. The superordinate goal of the current reform in science education worldwide is the induction of a switch from the currently dominating lower-order cognitive skills (LOCS)/algorithmic teaching to

HOCS/evaluative teaching and, ultimately, *learning* (Zoller, 1993, 1997; Zoller & Tsaparlis, 1997; Zoller et al., 1995).

2. Although the road to attain this goal is rocky, we can teach for HOCS learning and, ultimately, for interdisciplinary *transfer* (Solomon & Perkins, 1989), provided that appropriate teaching and corresponding assessment strategies, which prove to be successful via research, will be purposely and creatively implemented (Zoller, 1997, 2000).

3. Teachers are the key factor in making any educational reform work via persistent longitudinal, preplanned, and adequately imple- mented systemic process. Yet no matter how clear, and even agreed upon, an educational goal and teaching objectives might be, an all- important practical issue is how to translate these objectives (HOCS learning in our case) into science courses and curricula, teaching strategies, and, accordingly, alternative *assessment methodology* of the teaching/learning outcomes. In particular, critical thinking and decision making–oriented and/or interdisciplinary courses and pro- grams may help facilitate or enhance the transfer of HOCS (Zoller, 1993, 1997, 2000). Preservice and inservice teacher training and professional development programs should be designed and mod- eled not only to facilitate all of these points but also to empower teachers to do that and get it right.

4. Since there are differences between countries, communities, and societies, the HOCS-motivated reform in science education has had different expressions in different communities and contexts, nation- ally and internationally, concerning some emerging key issues and contextual factors.

In view of this, we have initiated and conducted within the unique science and mathematics teacher training program of our university the fol- lowing longitudinal, multidimensional, collaborative national and (compar- ative) international studies, focusing on a few selected key issues directly related to teachers' professional development and the quality of their teach- ing accordingly.

1. Assessment: LOCS versus HOCS-oriented exams and examination- type preferences of prospective science teachers as well as students' self-assessment

2. Critical thinking: Disposition toward critical thinking (DTCT)

All of these issues and related contextual factors (e.g., local culture, test- ing culture, teaching tradition, educational system, accepted social norms, and measure of success) are presented and thoroughly discussed, based on our collaborative corresponding research finding/results (in Israel, Italy, Greece, and the United States) within a case study format.

The Case of Assessment

One of the pressing topics within the realm of science education reform is that of assessment (Kulm & Malcolm, 1991). The latter is needed for facilitating the evaluation of students' performance in terms of HOCS (Zoller, 1993) and conceptual understanding (Smith, Blakeslee, & Anderson, 1993) and to serve as both an index of the teaching goals' attainment and a formative means for revising/remediating the teaching strategies and examination strategies accordingly (National Science Education Standards, 1996).

Despite the emerging consensus on the urgent need for education reform in general and science education reform in particular (AAAS, 1994; NSTA, 1993), a sharp contrast exists between the current visions of educational excellence and currently institutionalized patterns of educational practices (Raudenbush et al., 1993). *Examinations*, in particular, persistently remain the most regressive in this respect (e.g., Blinn, 1993; Romberg, Zarinnia, & Collis, 1991) and in dissonance with the current HOCS orientation.

In view of this, we have conducted several research studies within different contexts and locations that aimed at assessing the preferences of science students and their faculty concerning examination-types in the science disciplines and investigating the possible use of student self-assessment in HOCS-oriented science examinations.

Examination-Type Preferences

At the outset of our research a two-part Likert-type questionnaire—Types of Preferred Examinations (TOPE)—was developed (by the authors) and validated for assessing students' examination-type preferences (Zoller & Ben-Chaim, 1988). The first part consists of a list of nine types of examinations such as "an oral examination—all types of supporting material permitted"; "written examination in class—no supporting material (notes, textbooks, etc.) may be used and time duration is limited"; and "final project, or seminar paper as substitute for ordinary end-of-term exam." The second part of the TOPE questionnaire consists of six questions in which the students provide reasons for their preferences; the particular discipline(s) to which their response is related; examination types they have experienced; the effect, they believe, the type of exam has on their final grade in the course; and suggestion(s) concerning examinations and evaluation of student achievement. In each case, following the administration of the TOPE questionnaire a randomly selected sample of the research population was selected and interviewed using a structured interview format. The main purpose of the interview was to provide deeper insights into and a better understanding of the perceptions and rationales used by students for their exam-type preferences when their performance in science learning was being assessed. In addition, the faculty of the science and mathematics disciplines ranked their preferences regarding the examination types and were individually requested to respond briefly to three questions:

1. In your opinion, which types of exam are most preferred by your students?
2. Which types of exam would you prefer to use in your teaching and why?
3. What type(s) of examination do you actually use in your science (or mathematics) classes?

In a related representative cross-cultural study, we compared the examination-type preferences of college science students and their science faculty at a teacher training and a community college in Israel and the United States, respectively (Zoller, Ben-Chaim, & Kamm, 1997). The Israeli sample consisted of 152 prospective science teachers (46 males, 106 females) and 24 faculty staff at the department of science education. The American sample consisted of 57 students (22 males, 35 females) of general physics classes (primarily health related majors) and their science faculty ($N = 23$).

The conclusions of this study are based on the data given in Zoller, Ben-Chaim, and Kamm (1997), the excerpts from the Israeli and American students' responses in explaining their reasons for their examinations type preferences, and the excerpts from faculty responses in both countries concerning their preferences.

The following are selected typical excerpts from the Israeli students' responses concerning the reasons for their particular preferences:

> It is imperative to give priority to those types of exams which will reflect adequately the students' personal knowledge, real understanding and synthesis capacity. . . . Anxiety and pressure have a profound detrimental effect. . . . I never perform to the best of my abilities under such circumstances. . . .

> I prefer those kinds of exams which evaluate my deep understanding and thinking abilities, not my rote capacity and memorization of the ongoing knowledge. . . .

> The limited time in exams causes students to become "strained" and it fails them . . . unlimited time relieves the excitement and pressure and then one can express one's mastery of the material. . . .

> There is no sense in taking "closed book" exams; in real life in the future we are going always to use available material, and the time constraints for

solving a problem are not as those (always time pressure) as in the traditional exams. . . .

The following are selected excerpts from the American students' typical responses represent the reasons for their preferences:

I don't think you study adequately when you have take home exams or open book exams. . . .

I prefer the options that allow the usage of materials and those of unlimited time because they are low-pressure, more enjoyable, and contain an inherent potential for higher grades. . . .

For I (written exam in class; time unlimited; any materials allowed), the student wouldn't have to learn anything. With unlimited time anyone can find the answers. . . .

A written exam in class; time limited; no materials allowed seems the most strict and would prepare students for a university. This is my preference since one must be forced to learn the material. . . .

I have never had an easy "take home" test. . . . However, I've learned a great deal from them. . . .

The following are selected representative excerpts from faculty responses in both countries concerning their preferences:

"A—in senior level courses [Exam type A: Final project/seminar work]; I— in O-level science classes [Exam type 1: Written exam in class; time unlimited, all supporting material allowed]; E—for 1 or 2 exams during the term [Exam type E: Oral; in groups of 2–3; no material may be used]; D—for junior/senior level courses . . . . [Exam type D: Oral; individually, all supporting material may be used]" *(Amer.)*

For different courses different examination types. . . . *(Isr.)*

The examination type is course dependent . . . . In mandatory courses a written exam is preferred; an oral exam may cause a too high anxiety for some students . . . . *(Isr.)*

Hence, the findings of the comparative study suggest that:

1. College science students in Israel prefer (by far) the nontraditional examinations in which the emphasis is on thoroughness and understanding, thinking and analyzing and that the use of any relevant

material during the examination be allowed ("open books") and the time duration be practically unlimited. The HOCS-orientation in this examination type preference is clearly implied.

2. This preference is significantly higher for female compared with male students, whereas the oral examinations of all types are the least preferred type, and actually are rejected, significantly more by female students (see also Zoller & Ben-Chaim, 1990).

3. College science students in the US prefer most (mean of 3.14 on a scale range of 1–4) the nontraditional examination I (written; no time limitation; supporting material allowed) and slightly less, (mean of 3.02), the exam type H (written exam in class, time limited, all supporting material allowed). A significantly higher number of US students (mean of 2.98) preferred the traditional exam G (written exam in class; time limited, no supporting material allowed), compared to their Israeli counterparts (mean of 2.21).

4. In contrast to the results obtained in Israel, no gender differences were found concerning the most preferred examination type by the American college science students. The oral examinations were the least preferred, and actually rejected (as in Israel), significantly more for C and D (oral, individual) by female students.

5. There exists a significant gap, more so in Israel than in the United States, between the examination type preferences of college science students and their science faculty concerning B (take-home exam) and I, the most preferred by the Israelis, and I and H (written, time limited, supporting material allowed), the most preferred by the American students.

6. The traditional examination (G) is the most and second preferred examination type by the American and the Israeli science faculties, respectively, significantly more than their students' preference, that is, $\chi = 2.98$ (ranked third) and 2.21 (ranked fifth), respectively.

The last two findings are, probably, the most important with respect to the current efforts to reform science education: If, indeed, the development of students' HOCS and the attainment of conceptual understanding by students in the science teaching-learning process constitute core goals, then without being aware of the existing gap between the "examination-type profile" of students and faculty and the taking of appropriate action accordingly, including the active participation of students in this process, the success of the reform is questionable.

Student Self-Assessment

Self-assessment refers to the process of actively monitoring one's own progress in learning and understanding and of examining one's own content knowledge, processes, and attitudes (Kenney & Silver, 1993). It is agreed that student self-assessment of HOCS science examinations is in line with the ever-increasing call for integration of assessment and learning in teaching (Conway et al., 1995; Zoller, 1993, 1995). However, given the current state of affairs, student self-assessment of HOCS examinations may constitute a problem and raise questions. The immediate emerging questions (concerning both formative and summative assessments as an integral part of the evaluation system) are the following:

- Can the students do that?
- Are they confident in doing that?
- Is their assessment compatible with that of their teachers?

In our research (Zoller & Ben-Chaim, 1998) an attempt was made to investigate the state of the art concerning these questions, which turned to be the guiding research questions.

A specially designed Self-Assessment Questionnaire of High-Order Cognitive Skills Capability (SAQHOCS) has been developed (by the authors) and validated for obtaining empirical data in response to the guiding research questions (Zoller & Ben-Chaim, 1998). The questionnaire constituted a mixed bag of HOCS, interdisciplinary science-technology-environment-society (STES)-oriented (Zoller, 1990) and mathematics questions and Likert-type questions, with respect to students' confidence. The respondents were requested to self-assess and grade themselves and to provide criteria for their grading.

In addition, in order to obtain an in-depth insight into students' views concerning self-assessment and its application within science teaching and learning, structured interviews with a stratified sample (mainly freshmen prospective biology teachers) of the research population were planned to follow the administration of the SAQHOCSC. A five-item Students' Views On Self-Assessment Questionnaire (SVOSAQ) was constructed and used for this purpose.

The SAQHOCS questionnaire was administered to 71 prospective biology teachers (48F and 23M) enrolled in a 4-year Israeli college teacher training program. This was followed by conducting the interviews with 7 (4F and 3M) students. The prevailing teaching and "testing culture" in most of the mandatory science courses, mainly with respect to final exams in the first 2 years of study, has been LOCS-oriented, in accord with the dominant "test culture" in science teaching worldwide.

Our main conclusions are based on the results summarized in Table 12.1, which present the student self-assessment versus teacher assessment (0–100 scale) of students' responses to the exponential growth question, on the analyses of the students' appraisal of their capability and confidence in self-assessment and assessment of peers (in relation to Question 4 on

Table 12.1.

Student Self-Assessment vs. Teacher Assessment (0–100 scale) of Students' Solutions to the Exponential Growth Question (Question 2)

	Correct Solution	Naïve "Correct"	Naïve "Incorrect"	Awareness of Exponential Growth[a]	Attempted Incorrect or Irrelevant	No Response	χ
N	2	33	8	11	15	2	
Self-Marking Range	100	100 (21) 70–90 (12)	100	75–80[b]	60–100	0 (1) 40 (1)	77.8[c]
Teacher Marking	100	65	50	70	35	0	56.9

[a] These students were aware of the fact that exponential growth is the issue here and that an appropriate "formula," to be applied, exists. They wrote, however, that they could solve the problem because they had forgotten the formula.

[b] Students noted that had they remembered the formula, they would give themselves 100.

[c] Under 50 (14%); 55–70 (10%); 75–85 (20%); 90–95 (10%); 100 (46%).

SAQHOCS), and on the excerpts of students' responses to the questions during the interviews of the sample of seven students.

The following are selected notes and written excerpts of students regarding the criteria they used (Question 6 on SAQHOCS). Only 39 (55%) out of the 71 students provided two criteria (as required in Question 6), that they used in their self-assessment of Question 4; 12 (17%) provided only one criterion, and 20 (28%) provided none, although they did scored themselves. Whether these findings point to students' difficulty in defining criteria or in actually using criteria within the process of self-assessment is not clear. A substantial number of students who did provide criteria tried to rationalize the relative *low* marks (i.e., 60, 70, and even 80) they assigned to themselves by providing criteria in the "negative" direction, for example, "lack of information"; "there is a central idea, but it is not sufficiently detailed"; "I do not have knowledge and understanding on a high level"; "no precision." On the other hand, a great number of students rationalized their self-scoring by providing "positive" criteria, for example, "completeness of answer"; "feeling of confidence in the answer"; "understanding in the subject"; "investment in the answer."

The following are selected notes and excerpts of the interviewed students regarding their attitude to the idea and practice of self-assessment. Four students of the seven interviewed were against the idea of student self-assessment and its application within the learning process; three were in favor. In their own words:

Student 1: It is not a good idea. A student cannot evaluate himself alone. The teacher has additional considerations in the assessment which do not exist when the assessment is being done by the pupil. . . . A teacher can evaluate new ideas of the pupil; the pupil cannot . . . people wouldn't be objective unless they know that they are under the teacher's control. . . .

Student 2: It is not a good idea . . . people wouldn't be objective . . . (I, yes) . . . It is an unusable idea. It is not good in a competitive system, where people are afraid to fail. It is "built" for more mature people . . . perhaps in Master and Ph.D. programs. . . .

Student 3: Not a good idea . . . with some people it is good. Others will be unsuccessful in doing that. For me, for example, it is not good since I under-appreciate myself. It is impossible to use this idea, because people are not trustful. . . .

Student 4: There is an educational meaning in the idea, but it is unusable for obtaining evaluation on students: most people are objective, but each person's objectivity is different; there are those who are more strict, and those who are more flexible. . . .

For this group of interviewees, the idea of self-assessment is not a good one and is not usable, mainly because of lack of objectivity, lack of deep understanding and maturity on the part of the students, impossibility of ranking students, and differences between people in the ways they assess. Significantly, the learning aspects of self-assessment is not even mentioned by the student interviewees.

Three students of the seven interviewed considered self-assessment to be a good idea:

Student 5: It is a good idea, but . . . with reservation . . . you cannot rely on all people. For me, it is a good idea, since everybody knows precisely her/his *level of knowledge*. . . .

Student 6: A good idea, since every person knows precisely his *level of knowledge*. The idea may be tried out in all courses at all levels. . . . As far as my class is concerned, one can rely on the pupils. . . .

Student 7: The idea is good only to a certain extent. My learning is thorougher when I check; I learn from my mistakes. There are subject matters in which a pupil cannot assess herself . . . it depends on the learned subject. . . . Post [self-assessment] teacher control is mandatory.

It is clear that even those who are in favor of the application of self-assessment are quite reserved. Although they do believe that "everybody *knows* her/his level of *knowledge*," still, like their peers who were not in favor of self-assessment (in terms of grades), objectivity is their main concern, and they further believe that the method can be applied only when the questions have a correct/incorrect answer. In other words: Self-assessment is a good idea as long as it is applied for LOCS-type questions and exercises, but definitely not for HOCS examinations.

Based on this, it can thus be concluded that there is a gap between student self-assessment (overestimation) and their teacher assessment, as far as HOCS exam questions in the sciences are concerned, which means that there is a problem. This gap, in terms of grading/marking, appears to be a consequence of three factors: (1) the prevailing of the LOCS-orientation in contemporary science teaching; (2) the use of different criteria—mainly "correct/incorrect" (LOCS level) on the part of the students in their self-assessment compared with the level of thinking (i.e., "HOCS vs. LOCS") on the part of the teachers (the authors) in this study; and (3) the existence of a total separation between examinations/performance assessment (the teachers' domain) and learning, HOCS learning in particular, which as yet is not conceived by students to be within their sole domain and responsibility. Yet the results and the analysis of the data collected suggest that the study students were capable of self-assessment and felt reasonably confident in doing that, with respect to LOCS exam questions, the type with which they are most familiar and experienced. With respect to HOCS-type exam questions, the potential for self-assessment to become an integral part of the assessment process is there, but still there is a long way to go and purposed HOCS-oriented teaching and learning work, including the practicing of self-assessment, to be done.

In a complementary study (Zoller, Fastow, Lubezky, & Tsaparlis, 1999), prospective science teachers in a college teacher training program in Israel and science majors in Greece self-graded themselves, as did their professors, on a midterm take-home exam in a freshman chemistry course. The research methodology was essentially the same as that previously described (Zoller & Ben-Chaim, 1998), and the findings corroborated those of the previous study. Thus, the results of the Israeli students' prospective science teachers' evaluation of their capability and confidence in self-assessment and assessment of peers indicate that they believe that "to a reasonable extent" they are *capable* of self-assessment ($\chi = 2.84$ on a 4–1 scale; S.D. = 0.70) and feel *confident* in this process ($\chi = 2.95$ on a 4–1 scale; S.D. = 0.72). Similarly, they believe in their capability of assessing their peers and have confidence in doing so ($\chi = 2.68$; S.D. = 0.69). These findings are rather encouraging.

Nevertheless, a new finding in the complementary study of Zoller, Fastow, Lubezky, and Tsaparlis (1999) revealed that the gaps between the students' self-grading and their professors' grading of LOCS questions are fairly small and statistically nonsignificant, whereas the gaps in the grading of HOCS questions are relatively large. We conclude that students' self-assessment of chemistry exam questions requiring LOCS is compatible with that of their professors, but that requiring HOCS is not. Hence, persistent and purposeful HOCS-oriented chemistry teaching and learning work needs to be done. Since the development of students' HOCS capability is a major objective in the reform of science and chemistry education, HOCS-oriented teaching and learning strategies should become the focus of the teaching-learning process. Students self-assessment that is consonant with the HOCS orientation should, therefore, become routine in chemistry and other science teaching. It *should* be done and *can* be done. However, for students to become capable and confident in self-assessment they should be encouraged and guided by capable, enthusiastic HOCS teachers to become autonomous learners.

The Case of Critical Thinking

Terms such as critical thinking skills, scientific thinking, reasoning skills, logical thinking and critical thinking (CT), have been used, quite often interchangeably, in the professional literature. In the context of our research and teaching in science education, our concept of CT is of purposeful inquiry-oriented consistent rational, logical, reflective, and consequential evaluative thinking, in terms of what to accept or reject and what to believe in; this is followed by a decision what to do or not to do about it (Ennis, 1989; Facione, 1990; Perkins, Jay, & Tishman, 1993; Zoller, 1993, 2000). Then one should act accordingly and take responsibility for the decision made (Zoller, 1993, 1997). However, as valuable as this may be, not only are specific cognitive skills per se important, but learners must also be *willing*, not just *able*, to think critically. Consistent willingness, motivation, inclination, and a drive to be engaged in critical thinking while reflecting on significant issues, making decisions and solving problems, is the essence of a disposition toward critical thinking (DTCT) (Facione et al., 1995; Facione, Facione, & Giancarlo, 1997). A student's/learner's DTCT is, most probably, a necessary precondition for critical thinking and constitutes an integral part of the critical thinking capability. Hence, while people may have the skill(s) to think critically, unless this is demanded of them by some external force, they may not get involved in such activity. Such people do not have a strong disposition toward critical thinking; they are not internally motivated to think critically (Facione et al., 1997).

    Given (1) the increasing importance of teaching and learning thinking skills in science education worldwide (Mattheis et al., 1992; Zoller, 1993); (2) the apparently close interrelationships between critical thinking—a major component of HOCS—and the disposition toward critical thinking (Kuhn, 1999); and (3) the valuable information to science teachers about their students' cognitive/affective strengths and weaknesses which may be obtained from systematically gathered and reliably evaluated data, we have undertaken research concerning the strength of this disposition within the context of contemporary science education.

For this purpose, data regarding the DTCT of university students from Israel, Italy, Greece, and the United States were compared.

The California Critical Thinking Disposition Inventory (CCTDI) (Facione & Facione, 1992) was used as the research instrument for assessing the students' DTCT. This instrument, which is composed of seven positive aspects of the disposition toward critical thinking, was designed to measure the overall disposition profile of students. Accordingly, the CCTDI is composed of seven subscales: The Truth-seeking scale (the T-scale), the Open-mindedness scale (the O-scale), the Analyticity scale (the A-scale), the Systematicity scale (the S-scale), the CT Self-Confidence (the C-scale), the Inquisitiveness scale (the I-scale), and the Maturity scale (the M-scale). The conceptual descriptions of the subscales and the development and validation process of the CCTDI are fully described in Facione and Facione (1992), Facione et al. (1994), and Facione et al. (1995).

The CCTDI questionnaire was administered to university students—pre-service prospective science teachers in a teacher training program—in Israel ($N = 60$ sophomore and junior students, biology majors), in Italy ($N = 42$ sophomore students, in an environmental program) and in Greece ($N = 73$ freshman students, science majors). In addition, we refer to the relevant data obtained in the United States ($N = 537$, private university freshman students) which is given in the manual of the CCTDI (Facione & Facione, 1992).

The CCTDI reports scores ranging from 10 to 60 on each of its seven scales. The "bottom line" concerning scoring is as follows: A score of 40 or over indicates confirmation of the characteristic. A score of 30 or less indicates the opposite, that is, a disinclination or hostility toward that same characteristic. A score of 31–39 indicates ambivalence (Facione & Facione, 1992). The CCTDI reports a total score, which is the sum of its seven scale scores, ranging from 70 to 420. A total score of 280 or higher indicates a positive overall disposition toward critical thinking, whereas a total score of 210 or lower indicates a negative disposition regarding critical thinking.

The DTCT profiles of the Israeli, Italian, Greek, and U.S. university students are presented graphically in Figure 12.1, followed by the percentages of students who scored over 40 on each of the subscales, for the sake of comparison. Additional detailed data regarding the Israeli and the Italian samples are presented in Zoller et al. (2000). As can be seen in Figure 12.1[AR12.1], significant differences exist between the different national university students on almost each of the seven subscales of the CCTDI.

The scores of the T-scale, much below 40 for each of the student populations, suggest that the students are not inclined or eager to seek for the "truth" (or the "last word") in a given context. This is more so for the Israeli students (mean, 34.00) than for the Italian (mean, 36.76) and the Greek (mean, 36.60) students.

Most striking are the differences found between the university students

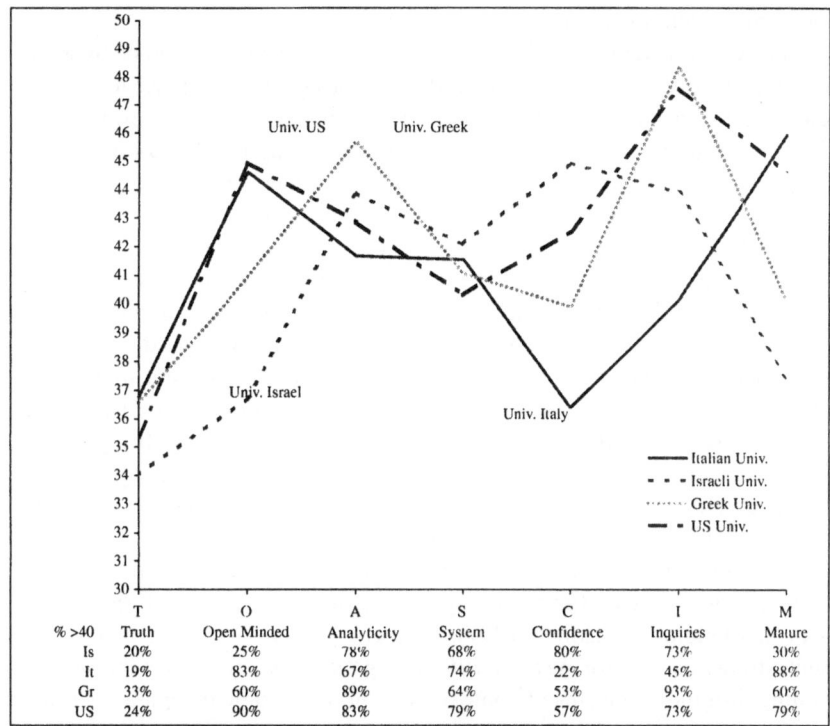

% >40	T Truth	O Open Minded	A Analyticity	S System	C Confidence	I Inquiries	M Mature
Is	20%	25%	78%	68%	80%	73%	30%
It	19%	83%	67%	74%	22%	45%	88%
Gr	33%	60%	89%	64%	53%	93%	60%
US	24%	90%	83%	79%	57%	73%	79%

Figure 12.1. Israeli, Italian, Greek, and American university student subscale-based profiles of disposition toward critical thinking and percentages of Israeli (Is), Italian (It), Greek (Gr), and American (US) students who scored over 40 for each subscale of the CCTDI.

of the four countries on the O-scale, C-scale, and M-scale: 25 percent of the Israeli versus 60–90 percent of the students of the other countries scored over 40 on the first subscale; 80 percent of the Israeli versus 22–57 percent, respectively, on the second subscale; and 30 percent of the Israeli versus 60–88 percent of the others on the third subscale (Figure 12.1). According to the criteria/categorization of the CCTDI developers, the Israeli students emerged as weak on the O-scale and M-scale and as strongly affirmative on the C-scale, compared with almost the opposite for the students from the other three countries. Given the nature of the items belong to those subscales, it is suggested that the differences found between the student populations of the four countries may reflect sociocultural differences, as well as a different instruction and learning "culture" between them.

Nevertheless, the total scores—283.11, 287.05, 292.63, and 298.22, for the Israeli, Italian, Greek, and U.S. samples, respectively, indicate a positive overall disposition toward CT for these populations.

Data from the respective Israeli and the Italian high school students indicate similar patterns and direction of the differences to those found between their university student samples (Zoller et al., 2000). Thus, it could be hypothesized that the overall disposition toward CT in the context of science teaching is determined at the high school level, without much later change or development. This, of course, has a profound implication concerning the development—via purposed teaching—of critical thinking in science students whether or not they are aiming at, or eventually will, become science teachers.

Analyses of the CCTDI scores by its subscales and by profiles, including the percentage of students who scored over 40 on specific aspects of CCTDI, might provide high school teachers, university faculty, and designers of preservice science teachers programs with data of relevance to their instructional strategies. They can adjust their teaching and assessment methods to reflect a greater or lesser emphasis on a given dispositional attribute. In particular, the results on the T-scale challenge teachers/faculty with bringing students to value open inquiry, attend to reasons, and move by relevant evidence. This is in consonance with the fostering/developing of students' CT within HOCS learning (Zoller, 2000).

It is quite obvious that educational goals shape the curriculum as well as the teaching strategies for achieving these goals and, in accord, the assessment methodologies (Zoller, 1997). Furthermore, new teaching and assessment strategies require new or *different* preservice and inservice teacher training programs in terms of context; content; pedagogical content; learning environment; teaching strategies; particular skills to be fostered, nurtured, and developed; faculty-student interactions; and more. All of this, in turn, affects professional development in science teacher education. In the context of the current HOCS-motivated reform in science education, HOCS-oriented assessment of student performance and the fostering of students' critical thinking constitute key issues that are directly related to teachers' professional development. These issues and selected related contextual factors are discussed next in view of our corresponding research findings/results in Israel, Greece, Italy, and the United States and their reflection in the evolving model of and current trends in teaching and assessment strategies in our university's science teacher educational program.

HOCS-Oriented Models of Science Teacher Education—Research into Practice

In the recent zeal to improve science teaching via HOCS learning to all (Zoller, 1997, 2000), teachers' professional development is viewed as central to this educational reform (NCTAF, 1996). Yet although the call for better professional development is unanimous, there are many who believe that it should be targeted toward improving student achievement (Dezier, 1998). From a research perspective, the problem is twofold: First, establishing a clear link between teachers' professional development and improved student learning is rather complicated; it requires substantial research and evaluation that carefully account for the specific contribution that each factor makes to the desired outcome (Guskey & Sparks, 1996). Second, relatively little research exists that addresses this connection(s) directly (Loucks-Horsley & Matsumoto, 1999). Thus, although much more research is needed in order to better understand the intricate connections between what, how, and levels of cognition that teachers learn and student learning, it is apparent that teacher learning and their professional development in preservice science education are critical in helping instruction to move beyond the mechanistic implementation of what has been experienced and learned to improve student learning.

We would argue that (1) the essence of the current reform in science education is a *LOCS teaching—HOCS learning shift* focusing on critical thinking (Zoller, 1993, 1997, 2000); (2) such a paradigm shift requires different conceptualization and goals of the teaching-learning process which, in turn, requires different models of teaching (Zoller, 1993, 1997), assessment (Zoller, 1993, 1997; Zoller & Ben-Chaim, 1998) and science teacher education (NRC, 1996) with consequential results, effects, and impact on the professional development and quality of teaching of the prospective teachers; (3) professional development based on systemic approach, goal(s) orientation, student thinking focus, and curriculum (not syllabus) is crucial in science teacher education (Loucks-Horsley & Matsumoto, 1999); (4) the LOCS to HOCS shift should be research-based (and driven), and the reform in science teacher education must match similar efforts in school reform (Simmons et al., 2000; Zoller, 1997, 2000); (5) we, science educators, must bring congruence to what we value and what we practice best as we continue to separate theory and research finding from practice; (6) cultural contexts of and beliefs about learning as well as meaning making in these contexts are crucial factors for understanding similarities and differences in teaching, learning, and their outcomes (Bempechat & Drago-Severson, 1999); and (7) successful HOCS-oriented models of science teacher education are (or will be) only those in which relevant educational research has been (or would be) translated into *practice and action* in terms of HOCS-oriented teaching and assessment strategies (Zoller, 1997) and, accordingly, professional development strategies (Loucks-Horsley et al., 1998).

Our action research program, only selected parts of which—the case of assessment and the case of critical thinking—are here presented and discussed, not only involves key issues in science teacher professional development and their expression in different cultural/national and teaching-learning culture contexts, but also reflects our actual teaching/practices, as well as the implementation of the respective research finding within our university's continuously changing science teacher education program. This, we believe, is the "right" (and the most promising) way for ending up with a successful research-based HOCS-oriented model of science teacher education within which research findings and guiding theories are translated into practice and action.

Thus, with respect to assessment in the science teaching context, our research findings suggest that:

1. Regardless of cultural differences, science students (more so prospective science teachers) prefer the nontraditional HOCS-type examinations, whereas their science faculty prefer the traditional paper-and-pencil LOCS-oriented examination.
2. Despite being aware of their students' preferences, the science faculty is stuck to the teacher-proof traditional LOCS examinations.

Our experience and research tell us that if one teaches for HOCS learning and assesses student achievement via "HOCS examinations" then, although the road is rocky, HOCS learning and respective transfer are attainable (Zoller, 2000). Needless to say, HOCS-oriented professional development of science teachers is a precondition for their students' HOCS learning.

With respect to student self-assessment capacity within their (HOCS) evaluative thinking skills we find that science students in different countries believe that they are *capable* of self-assessment and assessing their peers, and they are *confident* in doing that. However, we have found that the gap between students' self-grading and their professor's grading of LOCS exam questions are fairly small (and statistically nonsignificant), whereas the gaps in the grading of HOCS exam questions are relatively large, overestimated by the students.

Obviously, as one could expect, there are some differences in these results between countries because of differences in cultures, contexts, examination/grading/learning cultures, and, of course, the consequences of these practices. The overall picture concerning the present state of affairs is essentially the same. Therefore, student self-assessment in the science disciplines, on the one hand, and both the teaching and using of alternative assessment methodologies in our teacher training programs, on the other hand, have become an integral part of our science teaching and education with all the implications involved. Those concerning teacher professional development are apparent.

As far as the HOC-critical thinking skill is concerned, we have limited ourselves in this chapter to the dispositions toward critical thinking issue only. Generally speaking, our research findings point on an overall positive result, that is, "confirmation" of science students' dispositions toward critical thinking. The differences in scoring on the CCTDI subscales of the science students surveyed in different countries may be accounted for by sociocultural-educational-normative differences. However, why the overall DTCT of science students (including preservice science teachers) is more or less unchangeable between high school and college (freshmen) is unexplainable as yet, and the association or correlation (or causation) relationship between science students' DTCT and CT is still an open question currently under investigation (Zoller, Ben-Chaim, & Lubezky, submitted 2001). Whatever the results and conclusions concerning these issues, the vitality of the integration/inclusion of these issues (DTCT and CT) and the related teaching and assessment strategies within contemporary and future HOCS-oriented professional development and science educational programs is clear: It is essential in order to ensure—via research into practice—the success of both HOCS-oriented models of science teacher education and science education.

Concluding Remarks: Professional Development in Science Teacher Education and TIMSS Achievement Results—Is There a Correlation?

A leading rationale for this book was formulated as follows: "We have not, as educators, done a comparative analysis of what issues and perspectives arise during professional development for the teachers who teach both the students who have been studied and reported on in the TIMSS and those who have not been included in that sample. . . . [T]he authors (of this book) will investigate the following issues: What are some of the contextual factors in professional development? What science teacher education models are being used globally? What are the measures of success?"

In the planning stage of this chapter, as one of several research-based responses to these questions we hypothesized that "scoring at the upper and lower ends of the TIMMS sample has, perhaps, something to do with those factors of science teachers' professional development associated not with the LOCS-levels in traditional science teaching, but with their HOCS-related dimensions which are so crucial for a successful LOCS-to-HOCS switch in science and mathematics education" (Zoller & Ben-Chaim, 1999).

If, indeed, a switch from LOCS teaching to HOCS learning is the essence of the current reform in science education, then the fostering and development of prospective science teachers' HOCS capacity should constitute a major component of, and be in consonance with, responsive professional development programs and science teaching in science teacher education and science education at all levels.

Moreover, the measure of success in these respects is clearly the extent and level of attainment of students' HOCS capacity, regardless of whether in the short term they are prospective science teachers or in the long term their students within contemporary and future science education. Teachers' professional development in the HOCS domain is likely (albeit it does not guarantee) to promote the development of students' HOCS capability. The relevance and importance of HOCS-oriented teaching and assessment strategies as well as critical thinking (and related aspects) in this context is apparent.

The Third International Mathematics and Science Study (TIMSS) is, indeed, the largest, most comprehensive, and most rigorous international study of schools and student achievement ever conducted, and its scope is unprecedented in the annals of education research (Takahira et al., 1998). Unfortunately, it basically assesses LOCS-oriented student achievement. Therefore, regardless of whether or not the success of professional development should be measured by the improvement of the student achievement (and the monies allocated to these programs), the contribution of HOCS-promoting science teacher education cannot be evaluated via the LOCS-oriented TIMSS scoring. Consequently, our initial hypothesis in this respect cannot be put into test.

On the other hand, our contention that the superordinate goal of HOCS learning is attainable, provided that one purposely, persistently, and committedly teaches for HOCS learning, is corroborated both by our, studies and the research of others, as well as by the TIMSS study videotapes of classroom practices released worldwide (Olson, 1999).

In conclusion: Although the road toward HOCS learning is rocky, this ultimate goal of the current reform in science education is within reach. It *can* be done and it *should* be done.

References

American Association for the Advancement of Science (AAAS). Project 2061 (1994). *Benchmarks for science literacy: Ready for use!* New York: Oxford University Press.

Bempechat, J., & Drago-Severson, E. (1999). Cross-national differences in academic achievement: Beyond etic conceptions of children understanding. *Review of Educational Research, 69*(3), 287–314.

Blinn, L.V. (1993). Coping with cheating. *Journal of College Science Teaching, 23*(3), 173–174.

Bybee, R.W., Ferrini-Mundy, J., & Loucks-Horsley, S. (1997). National standards and school science and mathematics. *School Science and Mathematics, 97*(7), 325–334.

Conway, P.F., Clark, C.M., & Ben-Peretz, M. (1995, August). *The good test: A mathematic tool.* Paper presented at the European Conference for Research on Learning and Instruction. Nijmegen, The Netherlands.

Dezier, T. (1998). *1999 national awards program for model professional development.* Washington, DC: Office of Educational Research and Improvement, U.S. Department of Education.

Ennis, R.H. (1989). Critical thinking and subject specificity: Clarification and needed research. *Educational Researcher, 18*(3), 4–10.

Facione, P.A. (1990). *Critical thinking: A statement of expert consensus for purposes of educational assessment and instruction.* Washington, DC: ERIC ED, 315–423.

Facine, P.A., & Facione, N.C. (1992). *The California critical thinking disposition inventory.* Millbrae, CA: The California Academic Press.

Facione, P.A., Facione, N.C., & Sanchez, C.A. (1994). Critical thinking disposition as a measure of competent clinical judgement: The development of the CCTDI. *Journal of Nursing Education, 33*(8), 345–350.

Facione, P.A., Giancarlo, C.A., Facione, N.C., & Gainen, K.J. (1995). The disposition toward critical thinking. *Journal of General Education, 44*(1), 1–25.

Facione, P.A., Facione, N.C., and Giancarlo, C.A. (1996). The motivation to think in working and learning. In E. Jones (Ed.), Preparing Competent College Graduates: Setting New and Higher Expectations for Student Learning. In the series *New directions for higher education.* San Francisco, CA: Jossey-Bass Publications, pp. 67–79

Fraser-Abder, P. (Ed.). (in press). *Professional development in science teacher education: Local insights with lessons for the global community.*

Guskey, T., & Sparks, D. (1996). Exploring the relationship between staff development and improvements in student learning. *Journal of Staff Development, 17*(4), 34–38.

Kenney, P.A., & Silver, E.A. (1993). Student self-assessment in mathematics. In N.L. Webb & A.F. Coxford (Eds.), *Assessment mathematics classroom* (pp. 229–238). Reston, VA:) National Council of Teachers of Mathematics.

Kuhn, D. (1999). A developmental model of critical thinking. *Educational Researcher, 28*(2), 16–25, 46.

Kulm, G., & Malcolm, S.M. (Eds.). (1991). *Science assessment in the service of reform.* Washington, DC: American Association for the Advancement of Science.

Loucks-Horsley, S., Hewson, P., Love, N., & Stiles, K. (1998). *Designing professional development for teachers of science and mathematics.* Thousand Oaks, CA: Corwin Press.

Loucks-Horsley, S., & Matsumoto, C. (1999). Research on professional development for teachers of mathematics and science: The state of the scene. *School Science and Mathematics, 99*(5), 258–271.

Mattheis, F.E., Spooner, W.E., Coble, C.R., Takemura, S., Matsumoto, S., Matsumoto, K., & Yoshida, A. (1992). A study of the logical thinking skills and integrated process skills of junior high school students in North Carolina and Japan. *Science Education, 76*(2), 211–222.

National Commission on Teaching and America's Future (NCTAF). (1996). *What matters most: Teaching for America's future.* New York: Author.

National Research Council (NRC). (1996). *The national science education standards.* Washington, DC: National Academy Press.

National Science Education Standards. (1996). *Assessment in science education.* Washington, DC: National Academy of Science, National Academy Press.

National Science Teachers Association (NSTA). (1993). *Scope, sequence and coordination of secondary school science: The content core*, Washington, DC: NSTA. *Vol. I.*

Olson, S. (1999). Videotapes expose classroom faults. *Science, 283,* 1616–1617.

Perkins, D.N., Jay, E., & Tishman, S. (1993). Beyond abilities: A dispositional theory of thinking. *Merrill-Palmer Quarterly, 39*(1), 1–21.

Raudenbush, S.W., Rowan, B., & Cheong, Y.F. (1993). Higher order instructional goals in secondary schools: Class teacher and school influences. *American Educational Research Journal, 30*(3), 523–555.

Romberg, T.A., Zarinnia, E.A., & Collis, K.F. (1991). A new world of assessment in mathematics. In G. Kulm (Ed.), *Assessing higher order thinking in mathematics* (pp. 21–38). Washington, DC: American Association for the Advancement of Science.

Simmons, P.E., Yager, R., Craven, J., Brunkhorst, H., Gallager, J., Duggan-Haas, D., & McGlamery, S. (2000, April). *From theory to practice: How research in science teacher education informs and changes teacher education programs (part II).* Symposium presented at the National Association of Research in Science Teaching Conference, New Orleans.

Smith, E.L., Blakeslee, T.D., & Anderson, C.W. (1993). Teaching strategies associated with conceptual change learning in science. *Journal of Research in Science Teaching, 30*(2), 111–126.

Solomon, G., & Perkins, D.R. (1989). Rocky roads to transfer. Rethinking mechanisms of a neglected phenomenon. *Educational Psychologist, 24,* 113–142.

Takahira, S., Gonzales, P., Frase, M., & Salganik, L. H. (1998). *Pursuing excellence.* NCES 98–049 Office of Educational Research and Improvement, U.S. Department of Education.

Zoller, U. (1990). Environmental education and the university: The "problem" solving-decision-making act" within a critical system-thinking framework. *Higher Education in Europe, 15*(4), 5–14.

Zoller, U. (1991). Problem solving and problem solving paradox in decision-making-oriented environmental education. In S. Keiny & U. Zoller (Eds.), *Conceptual issues in environmental education* (pp. 71–87). New York: Peter Lang.

Zoller, U. (1993). Are lecture and learning compatible? Maybe for LOCS: Unlikely for HOCS. *Journal of Chemical Education, 70*(3), 195–197.

Zoller, U. (1995). Teaching, learning, evaluation and self-evaluation of HOCS in the process of learning chemistry. In R. M. Janink (Ed.), *Proceedings of the 3rd European Conference on Research in Chemical Education* (pp. 60–67). Lublin-Kanimierz, Poland.

Zoller, U. (1996). The development of students' HOCS—The key to progress in STES education. *Bulletin of Science Technology.* Vol. 1b(5–6), 268–272.

Zoller, U. (1997). The traditional-to-innovative switch in college science teaching: An illustrative case study on the reform trail. In M. W. Carpio (Ed.), *From traditional approaches toward innovation* (pp. 1–10). The SCST Monograph Series, Washington DC: NSTA.

Zoller, U. (2000). Teaching tomorrow's college science courses—are we getting it right? *Journal of College Science Teaching, 29*(6), 409–414.

Zoller, U., & Ben-Chaim, D. (1988). Interaction between examination-type, anxiety state, and academic achievement in college science: an action-oriented research. *Journal of Research in Science Teaching, 26*(1), 65–77.

Zoller, U., & Ben-Chaim, D. (1990). Gender differences in examination-type preferences, test anxiety, and academic achievements in college science education—A case study. *Science Education, 74*(6), 597–608.

Zoller, U., & Ben-Chaim, D. (1998). Student self-assessment in HOCS science examinations: Is there a problem? *Journal of Science Education and Technology, 7*(2), 135–147.

Zoller, U., & Ben-Chaim, D. (1999). Proposal for writing a chapter on "Key Issues and Contextual Factors in Professional Development of Preservice Science Teachers: Perspective from Israel, Greece, Italy and the US." Personal communication with the editor P. Fraser-Abder.

Zoller, U., Ben-Chaim, D., & Kamm, S.D. (1997). Examination-type preferences of college science students and (their) faculty in Israel and USA: A comparative study. *School Science and Mathematics, 97*, 3–12.

Zoller, U., Ben-Chaim, D., & Lubezky, A. (submitted). Relationship between teaching strategies and critical thinking skills and disposition of preservice prospective science teachers. *Journal of Research in Science Teaching*.

Zoller, U., Ben-Chaim, D., Ron, S., Pentimalli, R., & Borsese, A. (2000). The disposition toward critical thinking of high school and university science students: An inter-intra Israeli-Italian study. *International Journal of Science Education, 22*(6), 571–582.

Zoller, U., Nakhleh, M.B., Dori, J., Lubezky, A., & Tessier, B. (1995). Success on algorithmic and LOCS vs. conceptual chemistry exam questions. *Journal of Chemical Education, 72*(11), 987–989.

Zoller, U., & Tsaparlis, G. (1997). Higher-order cognitive skills and lower-order cognitive skills. The case of chemistry. *Research in Science Education, 27*(1), 117–130.

Zoller, U., Fastow, M., Lubezky, A., & Tsaparlis, G. (1999). Students' self-assessment in chemistry examinations requiring higher- and lower-order cognitive skills. *Journal of Chemical Education, 76*(1), 112–113.

Author Index

Subject Index